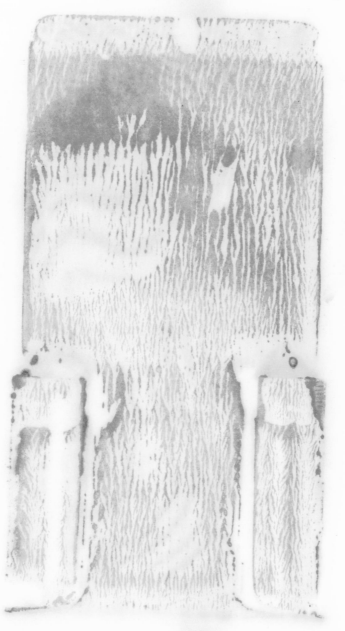

Principles and Practices of Incineration

ENVIRONMENTAL SCIENCE AND TECHNOLOGY

A Wiley-Interscience Series of Texts and Monographs

Edited by ROBERT L. METCALF, *University of Illinois*
JAMES N. PITTS, Jr., *University of California*

PRINCIPLES AND PRACTICES OF INCINERATION
Richard C. Corey

AN INTRODUCTION TO EXPERIMENTAL AEROBIOLOGY
Robert L. Dimmick

PRINCIPLES AND PRACTICES OF INCINERATION

EDITED BY

RICHARD C. COREY

Research Director
Pittsburgh Coal Research Center
Bureau of Mines
U. S. Department of the Interior

Wiley-Interscience

A DIVISION OF JOHN WILEY & SONS
NEW YORK · LONDON · SYDNEY · TORONTO

10   9   8   7   6   5   4   3   2   1

Library of Congress Catalog Card Number: 75-78479

SBN 471   17430   0

Printed in the United States of America

SERIES PREFACE

Environmental Sciences and Technology

The Environmental Sciences and Technology Series of Monographs, Textbooks, and Advances is devoted to the study of the quality of the environment and to the technology of its conservation. Environmental science therefore relates to the chemical, physical, and biological changes in the environment through contamination or modification, to the physical nature and biological behavior of air, water, soil, food, and waste as they are affected by man's agricultural, industrial, and social activities, and to the application of science and technology to the control and improvement of environmental quality.

The deterioration of environmental quality, which began when man first collected into villages and utilized fire, has existed as a serious problem since the industrial revolution. In the last half of the twentieth century, under the ever-increasing impacts of exponentially increasing population and of industrializing society, environmental contamination of air, water, soil, and food has become a threat to the continued existence of many plant and animal communities of the ecosystem and may ultimately threaten the very survival of the human race.

It seems clear that if we are to preserve for future generations some semblance of the biological order of the world of the past and hope to improve on the deteriorating standards of urban public health environmental science and technology must quickly come to play a dominant role in designing our social and industrial structure for tomorrow. Scientifically rigorous criteria of environmental quality must be developed. Based in part on these criteria, realistic standards must be established and our technological progress must be tailored to meet them. It is obvious that civilization will continue to require increasing amounts of fuel, transportation, industrial chemicals, fertilizers, pesticides, and countless other products and that it will continue to produce waste products of all descriptions. What is urgently needed is a total systems approach to modern civilization through which the pooled talents of scientists and engineers, in cooperation with social scientists and the medical profession, can be focused on the development of order and equilibrium to the presently disparate segments of the human environment. Most of the skills and tools that are needed are already in existence. Surely a technology that has created such manifold environmental problems is also capable of solving them. It is our hope that this Series in Environmental Sciences and Technology will not only serve to make this challenge more explicit to the established professional but that it also will help to stimulate the student toward the career opportunities in this vital area.

Robert L. Metcalf
James N. Pitts, Jr.

PREFACE

We are polluting our basic waste repositories—air, land, and water—to an alarming degree. In an era of the development of nuclear energy for the benefit of mankind, and the capability of extraterrestrial and underseas exploration, it is a paradox that we have only relatively primitive methods for disposing of our wastes, particularly combustible wastes.

Until recently we have cared little about contamination from such wastes as long as garbage and trash were removed. But unsanitary disposal practices, common in many parts of the United States, contribute significantly to air, land, and water pollution, breed rodents and insects, and deface property. Unless our waste-disposal practices improve these problems will become more serious because of the rapid increase in both the total and the per-capita rate of waste generation.

Incineration and sanitary landfill are presently the preferred methods for combustible-waste disposal. However, with the steady exhaustion of available disposal land within practical distances from population centers and the need for great care in selecting landfill sites (to prevent contamination of waters lying beneath the land), incineration may become the universal method of disposal, particularly in large communities. Moreover, incineration is becoming the preferred method for disposing of various combustible industrial wastes—gaseous, liquid, and solid—instead of discharging them to the atmosphere or lakes and rivers.

The purpose of this book is to bring together in one place discussions of the most important theoretical and practical aspects of incineration to serve as a guide for air-pollution-control officials and technologists, manufacturers and operators of incinerators, consultants, and students of environmental control. The examples given in some of the chapters will help with problems in incinerator design, operation, and performance.

Fortunately I was able to enlist the contributing authors, each of whom has a distinguished reputation in some aspect of incineration. Each author is solely responsible for the content and conclusions of his chapter. As editor I merely established the scope of the book and coordinated the authors' efforts. The personal opinions are those of the authors, not necessarily my own.

Improvements in incineration technology have become a moving target in recent years—more papers on research and development have been published during the last five years than during the previous 25 years. As new and significant findings come to light I shall welcome comments and suggestions from readers.

Richard C. Corey

Pittsburgh, Pennsylvania
July 1969

CONTENTS

Principles and Practices of Incineration

1

INTRODUCTION

F. R. Bowerman

Incineration is probably the second oldest form of waste disposal, dating from the time when man found that he could warm himself by burning the things he had hitherto dumped outside his cave. No longer did accumulating waste debris force him to leave his hunting ground for a new, less crowded, and probably less smelly area, but the problem of its complete disposal still remained. Nomadic groups, such as the Bedouins, with their constant treks from one location to another, have ignored the consequences of open waste dumps. Fixed communities cannot. Even today, all too often, stinking piles of debris smoulder on the outskirts of otherwise highly developed communities.

PURPOSES OF SOLID WASTE DISPOSAL

Examination of the basic functions and objectives of waste disposal helps place incineration in proper context with other means of waste disposal. Briefly the purposes of waste disposal are the following:

1. Removal of waste from premises to discourage vermin, to prevent odors, and to minimize fire hazards.

Group Vice President, Zurn Industries, Erie, Pennsylvania.

2. Deposition of waste in the earth or in the ocean, either as collected or as changed by incineration.

3. Salvage and reuse of waste as alternative to (2). Methods include utilization of waste heat; composting, or recovery of a mulch for soil improvement; pyrolyzation, or destructive distillation, for production of charcoal and combustible gas; recovery of steel from such articles as "tin" cans; and direct salvage of materials such as copper, aluminum, glass, even rags and cardboard, through sorting and separation.

Incineration affords two distinct advantages over other waste-disposal practices. First, it can be scaled to an apartment building or down to a single-family dwelling, accomplishing purpose (1) by reducing to a minimum the amount of waste to be collected from individual premises—and at the same time effectively sterilizing and oxidizing the putrescible waste to the point that it cannot support vermin nor emit odors. Second, the residue (or ash) left behind will require minimum space on or in the earth. Properly compacted, it will provide construction with a useful fill. This saving of space is particularly important to metropolitan areas that have limited disposal space within economical hauling distance. Incineration can extend the period of use of limited disposal sites by two to five times by using them only for unburned wastes [3]. Because of these considerable advantages incineration is regarded by many as the most promising answer to the increasingly difficult problem of waste disposal. Tempering this optimism is the serious need to provide incinerating devices that can operate without air pollution. Chapter 3 explores this factor in depth. To date incineration has the best potential for long-range disposal of unwanted combustible wastes.

EARLY HISTORY

One of man's earlier attempts to improve incineration resulted in a rather novel invention. A wagon, protected against fire by a coating of clay, was pulled by horses through medieval towns so that residents could throw burnable wastes onto the moving bonfire. Surprising proof of the old adage that "there is nothing new under the sun" was reincarnation of the idea during the period 1955 to 1960. Several innovators proposed that an enclosed mobile incinerator be drawn from one site to another for the burning of demolition wastes or household refuse.

In recording some of the earliest incinerators two American engineers, Hering and Greeley [2], credited Nottingham, England, with being the birthplace of municipal incineration in 1874, with Alfred Fryer as the contractor. British incinerators were called destructors. They were designed to burn garbage and rubbish without auxiliary fuel. They were so successful that by 1921 there were more than 200 plants in Great Britain. Combustion temperatures were high enough to generate steam and produce power. In the 1920s incineration was the only large-scale method of disposal used in England.

Hering and Greeley recorded that the first incinerator in the United States was built in 1885 on Governor's Island, New York Harbor, for the United States Government [2]. A municipal plant was built in Allegheny City, Pennsylvania, also in 1885. Both plants were called garbage crematories. They were designed to burn garbage with coal as auxiliary fuel. The furnaces were considered "low temperature." Frequent difficulties were caused by a temptation to lower the cost of operation by reducing the quantity of the fuel added [2]. The first furnace-type incinerator to prove popular in the United States was built in Des Moines, Iowa, in 1887. In it the garbage dropped from above onto one end of a large, horizontal grate, where it partially dried; it was then moved to the other end, where it burned just before the gases passed up the chimney. The auxiliary fuel was coal or oil. An adaptation of this furnace was built in Ellwood, Indiana, in 1893. It had an upper destructor chamber and a lower evaporation chamber; later a drying chamber was added above the main chamber. A popular variation on this design was first built by F. L. Decarie in 1901 at Minneapolis. In it the garbage was dried in a crate of steel pipes suspended over fire, with water circulated through the steel pipes. Steel water jackets also were used in place of firebrick in the furnace walls, but this kept the temperature too low for high-temperature combustion [2].

This concept of drying wet garbage prior to burning resulted in the continuing use of the basket-grate design for more than 50 years and the development of a later modification in which the wet refuse is supported on air-cooled, cast-iron fingers that rotate to release the dried material to a lower grate for final burning.

A small experimental rubbish-burning plant that incorporated the production of steam was built in New York City in 1903, followed by two full-scale plants in New York in 1906.

Hering and Greeley recorded that more than 200 municipal plants were operating in the United States by 1921. Many of the garbage crematories failed to operate satisfactorily, partly because of poor design, unskilled operation, and using too little auxiliary fuel.

A plant of English design with a capacity of 300 tons of mixed refuse per day was operating in Milwaukee by 1910. Up to that time all plants were hand charged, but mechanical-charging apparatus developed during the period 1910 to 1920 was installed in plants at Savannah, Atlanta, and Toronto, to name a few locations. All of the refuse incinerators built in the United States during this period were equipped with boilers for the production of steam. At Milwaukee the steam was used to generate electrical power for operating pumps for river flushing; at Westmount, Canada, for street lighting; at Savannah, for pumping part of the city's water supply.

The period from 1920 through 1950 saw the gradual development of automatic stoking grates in the United States. These evolved as four specific designs. One is a modification of the "beehive," or hemispherical, furnace, fed from directly above the center of the furnace to a circular hearth with a rotating, air-cooled, cast-iron rabble. The rotating arms of the rabble agitate the material, exposing new burning surfaces and allowing small particles and ash to fall through the grate into the ash hopper below. Two incinerator manufacturers developed similar designs, providing a highly competitive market for circular furnaces in the range of 100 to 200-tons/day capacity. Incinerating plants utilizing three or four such furnaces became almost an industry standard during the period 1940 to 1955 and still remain quite popular.

A second design is a modification of the multiple-hearth, mutual-assistance furnaces dating from about the 1920s. It has a sloping grate. The mechanization of the unit calls for partial rotation of the longitudinal sections of the grate starting from the top and working down so that the material is lifted and turned with new surfaces exposed as the wastes progress down the sloping hearth to the ash grates at the bottom of the slope. One of the advantages claimed for this configuration is that newly charged material is exposed to actively burning wastes on adjacent hearths because there is no wall separating the several charging doors that dump wastes onto the upper end of the hearth.

A third design has a traveling grate. In the long, narrow furnace an endless-belt hearth carries the wastes from the point of introduction to the discharge end; here the residue cascades into an ash hopper, usually a water bath. In this design the wastes are not tumbled or agitated while they burn.

A fourth design was transplanted from Europe to the United States when a rotary-kiln incinerator was constructed at Atlanta, Georgia, about 1925. This design has a sloping reciprocating hearth where the waste is dried before it is dropped into the rotary kiln. The ashes are discharged to an ash hopper at the lower end of the rotating kiln. These Atlanta kiln furnaces have had a long history of successful operation, incorporating utilization of the waste heat into the production of low-pressure steam for heating municipal buildings in downtown Atlanta. A similar installation in Los Angeles, California, was abandoned in the 1950s, partly because of excessive fly-ash discharge.

Table 1–1   Classification of Refuse Materials[a]

Refuse (Solid Wastes)	Composition		Source
Garbage	Wastes from the preparation, cooking, and serving of food Market refuse, waste from the handling, storage, and sale of produce and meat		Households, institutions, and commercial concerns such as hotels, stores, restaurants, markets, etc.
Rubbish	Combustible (primarily organic)	Paper, cardboard, cartons Wood, boxes, excelsior Plastics Rags, cloth, bedding Leather, rubber Grass, leaves, yard trimmings	
	Noncombustible (primarily inorganic)	Metals, tin cans, metal foils Dirt Stones, bricks, ceramics, crockery Glass, bottles Other mineral refuse	
Ashes	Residue from fires used for cooking and for heating buildings, cinders		
Bulky wastes	Large auto parts, tires Stoves, refrigerators, other large appliances Furniture, large crates Trees, branches, palm fronds, stumps, flotage		
Street refuse	Street sweepings, dirt Leaves Catch-basin dirt Contents of litter receptacles		Streets, sidewalks, alleys, vacant lots, etc.
Dead animals	Small animals: cats, dogs, poultry, etc. Large animals: horses, cows, etc.		
Abandoned vehicles	Automobiles, trucks		
Construction and demolition wastes	Lumber, roofing, and sheathing scraps Rubble, broken concrete, plaster, etc. Conduit, pipe, wire, insulation, etc.		
Industrial refuse	Solid wastes resulting from industrial processes and manufacturing operations, such as food-processing wastes, boiler house cinders, wood, plastic, and metal scraps and shavings, etc.		Factories, power plants, etc.
Special wastes	Hazardous wastes: pathological wastes, explosives, radioactive materials Security wastes: confidential documents, negotiable papers, etc.		Households, hospitals, institutions, stores, industry, etc.
Animal and agricultural wastes	Manures, crop residues		Farms, feed lots
Sewage-treatment residues	Coarse screenings, grit, septic-tank sludge, dewatered sludge		Sewage-treatment plants, septic tanks

[a] Data from [1].

All of these designs are currently used in the United States, with the choice of any particular design an individual matter since no clear-cut superiority appears to be demonstrated. The more recent history of incineration in the United States has seen the application of various types of scrubbing devices to minimize the amount of fly ash discharged to the atmosphere. Electrostatic precipitation, a method used in Europe for control of fly ash from incinerators, appears to be a likely method for the United States. Utilizing waste heat, a practice that fell into some disrepute from 1930 through about 1965, is again being considered in a more favorable light. The waste-heat-recovery system recently installed at an incinerator on the Atlantic seaboard to produce fresh water from ocean water may foretell greater emphasis on heat utilization.

THE NATURE OF SOLID WASTES

Incineration has its limitations; only the uninformed are willing to claim it as a total solution to a community's solid-waste-disposal problems. Roughly one-half of the solid wastes in a metropolitan area is burnable; the percentage originating from homes is about two-thirds by weight. A substantial amount of the unburnable solid waste in major metropolitan areas originates in industry, the demolition and reconstruction of structures, and the development of roads. For this reason the projected waste loadings must be carefully assessed whenever an incinerator is planned, so that the quantities to be burned may be fully understood. Unburnable solid waste from homes is usually collected together with garbage and

Table 1–2   Annual Quantities of Raw Refuse per Capita[a]

Class	Weight (lb)	Percent by Weight	Uncompacted Volume (yd³)	Percent by Volume	Specific Weight (lb/yd³)
Garbage	150	9.5	0.15	4.0	1000
Rubbish and all combustibles (except garbage)	1000	62.5	3.15	81.0	320
All noncombustibles	450	28.0	0.60	15.0	750
Total	1600	100	3.90	100	

[a] From Wegman, Leonard S., "Planning a New Incinerator," *Proceedings 1964 National Incinerator Conference*, ASME. Reproduced by permission of the American Public Works Association [1].

Table 1-3 Composition and Analysis of an Average Municipal Refuse from Studies Made by Purdue University[a]

Component	Percent of All Refuse by Weight	Moisture (percent by weight)	Analysis (percent dry weight)							Calorific Value (Btu/lb)
			Volatile Matter	Carbon	Hydrogen	Oxygen	Nitrogen	Sulfur	Noncombustibles[b]	
Rubbish, 64%										
Paper	42.0	10.2	84.6	43.4	5.8	44.3	0.3	0.20	6.0	7572
Wood	2.4	20.0	84.9	50.5	6.0	42.4	0.2	0.05	1.0	8613
Grass	4.0	65.0	—	43.3	6.0	41.7	2.2	0.05	6.8	7693
Brush	1.5	40.0	—	42.5	5.9	41.2	2.0	0.05	8.3	7900
Greens	1.5	62.0	70.3	40.3	5.6	39.0	2.0	0.05	13.0	7077
Leaves	5.0	50.0	—	40.5	6.0	45.1	0.2	0.05	8.2	7096
Leather	0.3	10.0	76.2	60.0	8.0	11.5	10.0	0.40	10.1	8850
Rubber	0.6	1.2	85.0	77.7	10.4	—	—	2.0	10.0	11330
Plastics	0.7	2.0	—	60.0	7.2	22.6	—	—	10.2	14368
Oils, paints	0.8	0.0	—	66.9	9.7	5.2	2.0	—	16.3	13400
Linoleum	0.1	2.1	65.8	48.1	5.3	18.7	0.1	0.40	27.4	8310
Rags	0.6	10.0	93.6	55.0	6.6	31.2	4.6	0.13	2.5	7652
Street sweepings	3.0	20.0	67.4	34.7	4.8	35.2	0.1	0.20	25.0	6000
Dirt	1.0	3.2	21.2	20.6	2.6	4.0	0.5	0.01	72.3	3790
Unclassified	0.5	4.0	—	16.6	2.5	18.4	0.05	0.05	62.5	3000
Food Wastes, 12%										
Garbage	10.0	72.0	53.3	45.0	6.4	28.8	3.3	0.52	16.0	8484
Fats	2.0	0.0	—	76.7	12.1	11.2	0	0	0	16700
Noncombustibles, 24%										
Metals	8.0	3.0	0.5	0.8	0.04	0.2	—	—	99.0	124
Glass and ceramics	6.0	2.0	0.4	0.6	0.03	0.1	—	—	99.3	65
Ashes	10.0	10.0	3.0	28.0	0.5	0.8	—	0.5	70.2	4172
Composite Refuse, as Received										
All refuse	100	20.7	—	28.0	3.5	22.4	0.33	0.16	24.9	6203

[a] Data from [1].
[b] Ash, metal, glass, and ceramics.

7

combustible rubbish from a single container or group of containers and transported in the same collection vehicle. Since this practice is convenient and economical for the homeowner, it probably will be continued, and incinerators will have to handle both burnable and nonburnable solid wastes in reasonable proportion. The percentage of burnable material in wastes is increasing because the trend in packaging in the United States is toward burnable materials such as paper and plastics. This change in the composition of wastes enhances the future use of incineration as a convenient means for reducing the volume of solid wastes that must be ultimately disposed of on or in the earth.

Tables 1–1 to 1–3, reproduced with the permission of the American Public Works Association [1], explain the nature of material to be considered for incineration; Table 1–1 shows the accepted standard definition and classification of solid wastes in the United States; Table 1–2, the percentage of waste classifications; and Table 1–3, a detailed breakdown of the composition of domestic refuse as determined by Purdue University.

REFERENCES

[1] American Public Works Association, Public Administration Service, *Refuse Collection Practice*, Third Edition, 1966.
[2] Hering, Rudolph, and S. A. Greeley, *Collection and Disposal of Municipal Refuse*, McGraw-Hill, New York, 1921.
[3] University of California, Technical Bulletin No. 5, Sanitary Engineering Research Project, *Municipal Incineration*, October 1951.

2

PRINCIPLES OF COMBUSTION

Arthur A. Orning

The combustion of fuels is among the oldest of engineering arts. The difficulty in reducing this art to a science is illustrated first of all by problems of definition. Combustion, in a more general area than that of interest here, has been defined as "any chemical process accompanied by the evolution of light and heat, commonly the union of substances with oxygen; hence, slower oxidation, as in the animal body." Since the objective of incineration is the conversion of waste materials—such as paper, wood, plastics, garbage, and other combustible refuse—to gaseous products and solid residues less voluminous than the original material, and since this should be accomplished as economically as possible, the special definition of combustion that is of interest here is a process of oxidation of substances with oxygen in air, commonly associated with vigorous evolution of light and heat.

Products of complete combustion are defined as carbon dioxide, water, sulfur dioxide, and nitrogen, generally the most highly oxidized state that is stable for each element of the fuel. This definition of complete combustion is somewhat arbitrary and incomplete. Some nitrogen is converted to oxides, particularly nitric oxide. Part of the sulfur is con-

---

Supervisory Physical Research Scientist, Pittsburgh Coal Research Center, Bureau of Mines, U.S. Department of the Interior, Pittsburgh, Pennsylvania.

verted to its trioxide rather than the dioxide. Metals such as iron and aluminum may be converted to their oxides. Neglecting oxygen consumption by such elements and the heat capacity of their combustion products, it is relatively easy to calculate the air required for complete combustion of a given fuel, the composition of the combustion products, and the temperature produced when a given portion of the heat released is held in the combustion products. The calculation is illustrated in the section on material and thermal balances.

In the special application to incineration, combustion is a high-temperature process. Heat released by combustion is partly stored in the combustion products and partly transferred by conduction, convection, and radiation both to the incinerator walls and to the incoming fuel. Heat transfer to incoming fuel is needed for ignition. Some aspects of ignition are discussed in relation to the management of fuel and air. Other aspects of ignition are discussed in relation to smoke and air pollution. The emission of air pollutants is discussed here only in relation to the principles of combustion.

MANAGEMENT OF FUEL AND AIR

The basic requirement for good combustion is that fuel and air must meet in proper proportions and under such conditions that prompt ignition and combustion occur. These requirements are relatively easy to meet with gaseous, liquid, or pulverized solid fuels. They are more difficult to meet with refuse materials because of the variability in the physical and chemical properties of these materials. Because of this variability and the prevalence of such materials as glass and metals (which makes it costly to process the refuse to produce a uniform fuel), the combustion is generally carried out in deep fuel beds.

It will be helpful to consider first an idealized system, such as a bed of double-screened coke, to understand what occurs in a bed of refuse material.

Depending on the relative movement of fuel and air, and the resultant flow of heat and combustion products, fuel beds may be classified [9] as illustrated later and as follows:

1. Underfeed beds in which fuel and air flow in the same direction.
2. Overfeed beds in which fuel and air flow in opposite directions.
3. Crossfeed beds in which fuel and air flow in directions at some angle to each other so that the heat needed for ignition neither flows directly against the air stream nor directly with the initial combustion products.

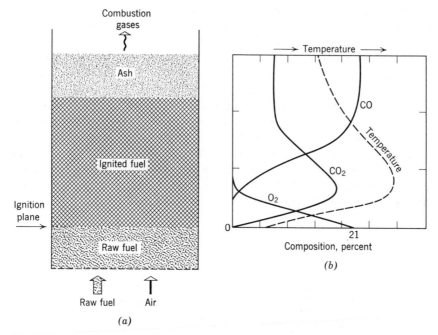

Figure 2–1 Idealized underfeed fuel bed (*a*), and relative distribution of temperature and products of combustion (*b*).

An idealized underfeed bed is shown in Figure 2–1. Flow of heat against the air stream produces a sharp temperature gradient at a level over the raw fuel designated as the *ignition plane*. Oxygen is rapidly consumed by the reaction

$$O_2 + C \rightarrow CO_2$$

as the gases move through the ignition plane and into the ignited fuel. As the oxygen is depleted the reaction

$$CO_2 + C \rightarrow 2CO$$

causes the carbon dioxide concentration to decrease as the carbon monoxide increases. This reaction absorbs heat so that the temperature falls as the gases flow toward the ash layer above the ignited fuel.

An overfeed fuel bed is shown in Figure 2–2. The incoming air passes first through the ash layer; heat recuperation favors higher temperatures and greater and more rapid conversion of carbon dioxide to the monoxide.

A crossfeed fuel bed is illustrated in Figure 2–3. As in underfeed burning, gas flow is from raw fuel into the ignited fuel, but heat flow for ignition does not flow directly against the gas stream.

The relative directions of fuel and gas flow, rather than absolute directions and velocities, are important in the classification of fuel beds. Batch burning in a pot with top ignition and air feed from the bottom would be classed as underfeed burning even though there is no fuel flow. Similarly, bottom ignition with air feed from the bottom would be overfeed burning. Horizontal fuel flow with air feed from the bottom, as on a moving grate, and corresponding to a 90° rotation of Figure 2–3, would be crossfeed burning. Front-end hopper feeding of a relatively fast moving grate approaches pure underfeed burning. Spreader-stoker firing to a deep bed on a moving grate approaches pure overfeed burning.

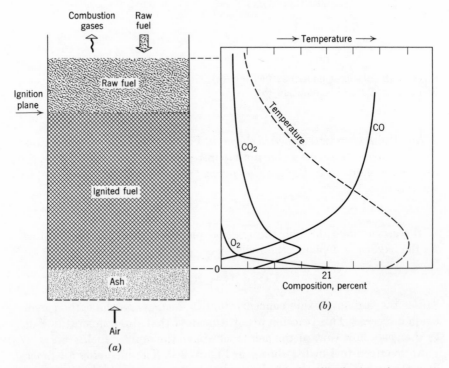

Figure 2–2 Idealized overfeed fuel bed (a), and relative distribution of temperature and products of combustion (b).

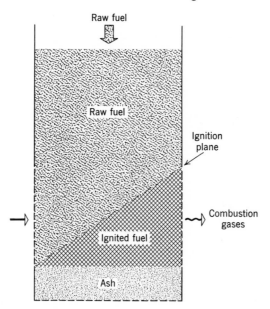

Figure 2–3 Idealized crossfeed fuel bed.

The rate of ignition may be measured in terms of the rate of advance of the ignition plane into the raw fuel. The plane of ignition may be defined as the plane within the fuel bed beyond which there is a sharp temperature gradient from that of the raw fuel to that of fully ignited fuel. This plane is most sharply defined for underfeed burning, in which heat must flow against the stream of incoming air; the oxygen in the gas stream is rapidly consumed so that no oxygen may remain after a travel distance of a few particle diameters within the bed and beyond the ignition plane. Ignition rates have been intensively studied for underfeed burning of coke [2]. A typical ignition rate curve is shown in Figure 2–4. The ignition rate at first increases rapidly with air-flow rate and exceeds the rate of burning of the ignited fuel. This is analogous to conditions at the head end of a moving-grate stoker. The ignition plane moves rapidly down through the bed. Distance from head end corresponds to increasing time after ignition of the top surface of the bed. The thickness of ignited fuel increases with distance, or time, until the ignition reaches the grate surface. The rate of ignition, as shown in Figure 2–4, passes through a maximum as the air-flow rate is increased. In the coke-burning tests the ignition rate became equal to the rate of burning of the ignited fuel at a point somewhat

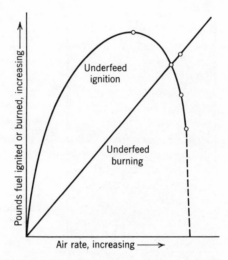

Figure 2-4 Typical curves for underfeed ignition and burning.

beyond the maximum ignition rate. Further increase in air-flow rate caused loss of ignition, instability, or generally unsatisfactory operation. It is partly for this reason that zoned draft may be used to limit air flow in the ignition zone of moving-grate stokers.

This discussion of ignition has been based on idealized beds of material such as double-screened coke that may be treated as a mixture of carbon and ash. Beds of refuse materials are far from ideal. Combustion air tends to flow through the more permeable portions of the bed while other portions are starved for air. Differences in moisture content in different parts of the bed tend to heighten this effect. Heat, needed for vaporization of water, is needed in addition to that needed for ignition. Steam flow from concentrations of material with a high moisture content also tends to disturb the air flow.

The lack of ideal conditions is particularly important in relation to the combustion of volatile matter. The volatile matter must receive oxygen in sufficient amount for complete combustion, at temperatures high enough to insure ignition. Heat flow into portions of the bed that are starved for air causes the release of volatile matter in gas streams that tend to be low in oxygen and too cool for prompt ignition even though oxygen was present in sufficient concentration.

Considerable amounts of carbon monoxide, as shown in Figures 2–1 and 2–2, together with volatile matter and other products of incomplete combustion, must receive oxygen either by a supply of overfire air or by mixing with oxygen-rich gas streams that escape the fuel bed. This over-fire mixing must be done under such conditions that the combustible gases will promptly ignite and burn. To do this is more an art than a science. Sufficient air is needed for complete combustion. Too much air, by reducing the temperature of the resulting mixture, will tend to hinder ignition of the combustible gases. Delayed mixing will permit cooling of the hot gas before the oxygen is supplied and again hinder ignition. Lazy mixing of fresh air with the hot gases will have the same effect as delayed mixing for streams that are further away from the fresh-air supply.

The management of fuel and air can take many forms. Good management can be illustrated best by summarizing conditions that tend to give good combustion in deep fuel beds. These are as follows:

1. Raw fuel should be fed in such manner as to give underfeed or cross-feed burning. Large masses of raw fuel should never be dumped on top of the fire.

2. As in underfeed and crossfeed burning, gas flow should be from raw fuel into ignited fuel.

3. Overfire air must be mixed with the hot gases flowing out of the fuel bed. This is best accomplished by such devices as overfire-air jets to give turbulent mixing close to intense-burning zones. Any streams of gas from colder portions of the fuel bed must be directed into zones of intense burning.

4. Additional mixing with high amounts of excess air that may be desired to give lower gas temperatures must be delayed so as not to hinder good combustion.

More novel modes of management of fuel and air are used in vortex [11] and fluidized-bed [1] combustion. In vortex combustion all of the air can be supplied as overfire air. The aerodynamics of the vortex is such that the air must follow a downward spiral to scrub the fuel bed and mix with rising volatile matter. It is only after the air is converted to hot, low-density combustion products that it can escape through the center of the vortex. In fluidized-bed combustion the refuse is fed into a turbulent bed of granular inert material. Through proper balance between heat input in fuel, heat loss to the surroundings, heat extraction from the bed, and enthalpy of combustion products leaving the bed, the temperature can be held at any desired level that gives good combustion. The temperature also is practically constant throughout a well-fluidized bed. Good heat transfer between hot granular material and incoming refuse causes prompt devolatilization and ignition of the refuse. The bed inventory must consist mostly of hot granular material and a small percentage of combustibles in order to operate with excess air. Temperature control requires a constant rate of heat input. The high variability in the heating value of refuse causes difficulty in control. The use of an auxiliary fuel is helpful. Tin cans tend to disintegrate. Some noncombustible materials, such as glass bottles, would tend to sink to the bottom of the bed, whereas others might remain suspended in the bed. Continuous extraction and reclassification of noncombustibles from a recycle stream is needed to maintain the right fluidization properties in the bed of inert granular material. Efficient dust removal from the flue gas is also needed because of the high carry-over of particulate matter from the fluidized bed.

MATERIAL BALANCES

Since temperature, as well as the contacting of fuel and air, is important in determining the progress of combustion, it is well to examine the effect of refuse composition on the air requirements, and the effect of excess air on the gas temperatures produced. Information necessary for such computation is the analysis of air, and the ultimate analysis and gross heating value of the refuse.

The composition of air is approximately constant, except for moisture content. The assumed composition on a basis of 70-percent relative humidity at 60°F is given in footnote (b) of Table 2–1.

Refuse composition varies considerably, depending on such things as the source (domestic, commercial, etc.), geographic area, and climatic conditions. Computations are illustrated for a particular refuse [6]. The same numerical procedures may be used for other refuse compositions.

Computations are presented first in considerable detail. It may not be necessary to carry as many significant decimal places as shown in the illustrations. The number of significant places was chosen to enable the reader to verify the computation and to judge the simplifications and approximations that he might choose to make.

Material-balance calculations are made in terms of atomic-weight units of elements in the fuel and moles of combustion products because of the simplifications that result. These simplifications appear because each atomic-weight unit contains the same number of atoms; and each mole of gas, under standard conditions of temperature and pressure, occupies the same volume. The last assumes ideal gas behavior, which is true within practical limits for air under atmospheric conditions as well as for combustion gases, excepting moisture condensate at temperatures below the dew point. Within these limits the gas composition in percent by volume is the same as mole percent.

Table 2–1 illustrates the calculation of air requirements and of the resulting gas composition. The second column of this table gives the

Table 2-1 Calculation of Air Requirement and Gas Composition at Zero Excess Air[a]

Component	Weight Percent	Atomic Weight	Atomic Weight Units	Moles of Oxygen Required	Combustion Product	Moles of Combustion Products			
						From Combustion	From Air[b]	Total	Percent
Carbon	29.83	12.011	2.484[e]	2.484[e]	Carbon dioxide	2.484	0.004[d]	2.488	15.5
Hydrogen	6.23[e]	1.008	6.181	1.545	Water	3.090	0.166[d]	3.256	20.3
Oxygen	43.45[e]	16.000	2.716	-1.358	—	—	—	—	—
Nitrogen	0.37	14.008	0.026	—	Nitrogen	0.013	10.283[d]	10.296	64.2
Sulfur	0.12	32.066	0.004	0.004	Sulfur dioxide	0.004	—	00.004	00.02
Ash and metal	20.00	—	—	0.052[f]	—	—	—	—	—
Total	100.00			2.727				16.044	100.0

Moles of air required per 100 lb fuel = 2.727/.2069 = 13.18

Pounds of air required per pound of refuse = 13.18(28.7)/100 = 3.78

Moles of air per mole of gas = 13.18/16.044 = 0.8215

Moles of gas produced per pounds of refuse = 16.044/100 = 0.16044

[a] Computation is based on 100 lbs of refuse.

[b] Assumed air composition, in volume fractions: carbon dioxide, 0.0003; nitrogen, 0.7802; oxygen, 0.2069; water, 0.0126. Assuming ideal gases, the volume fractions may be taken as mole fractions and are equal to the percentages by volume divided by 100. The composition as given is for rare gases included with the nitrogen and with moisture content corresponding to 70% relative humidity at 60°F. Air of this composition has a weight of 28.7 lb per mole of total gas.

[c] Per 100 lb of refuse.

[d] 13.18(0.0126) = 0.166, for instance.

[e] Includes hydrogen and oxygen from 20% moisture.

[f] An assumed value for partial burning of metals.

weight percent in the particular refuse for each element as identified in the first column. On the basis of 100 lb of refuse, the weight percents equal pounds of each element. Division by the atomic weights gives for each element the atomic-weight units shown in the fourth column. Each atomic-weight unit of carbon requires 1 mole of O_2 to produce 1 mole of CO_2. These quantities are shown in the fifth and seventh columns of the table. An atomic-weight unit of hydrogen requires $\frac{1}{4}$ mole of O_2 to produce $\frac{1}{2}$ mole of H_2O. Each atomic-weight unit of oxygen in the refuse represents a $\frac{1}{2}$ mole of O_2 that need not be supplied by air to form combustion products. It is assumed that the nitrogen of the refuse forms N_2 in combustion products. This neglects small percentages of nitrogen oxides that are found in combustion gases. The error is negligible in comparison to the percentages of other combustion products. Thus each atomic-weight unit produces $\frac{1}{2}$ mole of N_2. Like the carbon, each atomic-weight unit of sulfur requires a mole of O_2 to produce a mole of SO_2. Since the sulfur percentages are low and only a few percent of the SO_2 is oxidized to SO_3, the moles of SO_3 and the extra oxygen required for its formation can be neglected in determining gas composition and the oxygen requirement.

Note that the total oxygen requirement, 2.727 moles, includes a negative item corresponding to oxygen in the refuse. The air required at zero excess, 13.18 moles per 100 lb of refuse, is found through division of the O_2 requirement by the volume (or mole) fraction of oxygen in the air, as shown below the table.

The weight per mole of a mixed gas is easily calculated as the sum of volume fractions times molecular weights. Thus for air of the assumed composition (note b, Table 2–1)

$$0.0003(44.011) + 0.7802(28.016) + 0.2069(32) + 0.0126(18.016)$$
$$= 28.7 \text{ lb/mole.}$$

This permits immediate calculation of 3.78 lb air required per pound of refuse, as shown. The moles of combustion products derived from the air are found by multiplying the moles of air required by the volume fractions in air of CO_2, H_2O, and N_2. All the oxygen of the air, at zero excess, is included in the combustion products of the refuse. Since percent by volume is practically equal to mole percent, the composition of the product gas is obtained from the total moles for each gas as a percentage of the grand total.

The calculation of gas composition at various excess air levels is illustrated in Table 2–2. The computation is based for convenience on 100 moles of combustion products at zero excess air. Total moles per 100 moles of combustion products at zero excess air, given in column 3 for each excess air level, will be useful later in calculating thermal balances.

Table 2–2   Effect of Excess Air on Gas Composition[a]

Percent Excess Air	Moles Excess Air[c]	Total Moles	Gas Composition[b]				
			CO_2	O_2	N_2	H_2O	SO_2
0	0.00	100.00	15.50	0.00	64.20	20.30	0.02
			15.50	0.00	64.20	20.30	0.02
50	[d]41.08	141.08	15.51	8.50	96.23	[e]20.82	0.02
			11.0	6.0	68.2	14.8	0.01
125	102.69	202.69	15.53	21.24	144.31	21.59	0.02
			7.7	10.5	71.2	10.6	0.01

[a] Based on 100 moles of gas at zero excess air.

[b] Composition is given in moles in the first line and in percent in the second line.

[c] Percent of excess air $= \dfrac{100 \text{ (air supplied} - \text{theoretical air requirement)}}{\text{theoretical air requirement}}$.

[d] Excess-air requirement is calculated from the ratio of the theoretical air requirement to gas produced (from Table 2–1) as

$$\left(\frac{\text{moles of air}}{\text{moles of gas}} \right) (\text{percent of excess air}) = 0.8215(50) = 41.08$$

moles of excess air per 100 moles of gas produced at zero excess air.

[e] Calculation of the moles of product gas is illustrated for water at 50% excess air as

$$(\text{moles excess air}) \left(\frac{\text{moles of water}}{\text{moles of air}} \right) + (\text{moles of water per 100 moles of product gas}$$

at zero excess air) $= 41.08(0.0126) \times 20.30 = 20.82$.

The gas composition is given on a moist basis. Conventional gas analyses are generally on an approximate dry basis. Computation of percentages on a dry basis requires division by 100 minus the percent moisture and multiplication by 100. This computation is left to the reader. The percentages on the moist basis are appropriate to computation of thermal balances.

An approximate calculation of theoretical air requirements, satisfactory for many engineering purposes, can be based on the heating value of refuse and a generality that appears among fuels. Table 2–3 gives the pounds of air required to release 1000 Btu on complete combustion of the given fuel for a series of organic compounds containing carbon, hydrogen, and oxygen. The oil (glycol dipalmitate) and cellulose (the principal organic component of paper) are representative of major components of refuse materials. Other components (proteins, plastics, etc.) are not included in the table, but the same approximation probably applies with somewhat less precision. Figure 2–5 is drawn on the basis of an air requirement of

0.7 lb per 1000 Btu of heat released. If the gross heating value of the refuse is known, the pounds of air required per pound of refuse can be read directly from the figure for various excess air levels.

Either the ultimate analysis or the heating value of the refuse is needed to calculate air requirements. A modified Dulong formula is sometimes useful in estimating gross heating value from the fuel analysis:

$$\text{gross Btu per pound} = 145.4\ C + 620\left(H - \frac{O}{8}\right) + 41\ S,$$

where C, H, O, and S are the weight percentages in the refuse of carbon, hydrogen, oxygen, and sulfur, respectively. The term $(H - O/8)$ represents hydrogen in the fuel that is combined with oxygen as moisture

Table 2–3   Air Requirement per 1000 Btu for
Various Compounds

Compound	Formula	Pounds of Air per 1000 Btu
Methane	CH_4	0.73
Methyl alcohol	CH_3OH	0.67
Propane	C_3H_8	0.73
Propylene	C_3H_6	0.71
Benzene	C_6H_6	0.74
Glucose	$C_6H_{12}O_6$	0.69
Glycol dipalmitate	$(C_{15}H_{31}CO_2CH_2)_2$	0.74
Cellulose	$(C_6H_{10}O_5)_x$	0.68

as well as hydrogen and oxygen that are combined with the fuel in some other form. Insofar as hydrogen is generally present in amounts exceeding that of the equivalent oxygen, an amount of heat equal to the heat of formation of the water may be considered to have been released in the original formation of the oxygen-containing fuel.

An approximate calculation of excess air can be obtained from the product-gas analysis as expressed in percent by volume. The excess air is given as

$$\text{percent of excess air} = \frac{100(O_2)}{(20.69/78.02)(N_2) - (O_2)}.$$

The denominator of this expression represents the oxygen requirement as the difference between oxygen equivalent to the nitrogen supplied and the oxygen remaining as such. One hundred times the ratio of the oxygen remaining to the oxygen supplied is then the percent of excess air. This

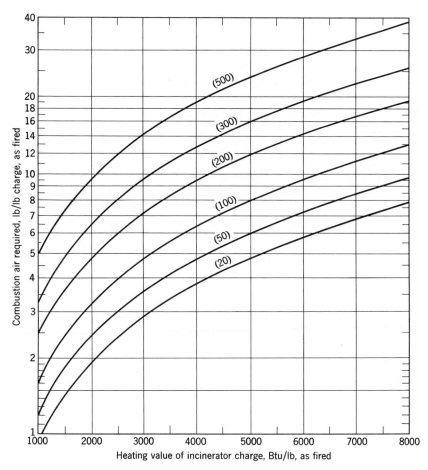

Figure 2–5 Air requirement versus heating value of refuse. Numbers in parentheses denote percent of excess air.

computation is approximate because of the sulfur and nitrogen content of the fuel.

Sulfur dioxide is included with carbon dioxide in gas analysis. These gases involve the same oxygen consumption per mole. Error due to sulfur content is limited by the small retention of sulfur in the ash and by the small conversion to sulfur trioxide instead of the dioxide. Error due to nitrogen in the fuel is also small because of the relatively large portion of the nitrogen that is supplied by the air. Error in the present illustration is less than that due to the limited precision of gas analyses. It may become significant with materials that are high in nitrogen, such as proteins.

THERMAL BALANCES

Computation of gas temperatures requires a knowledge of enthalpies. The enthalpy for a given amount of gas is the heat needed at constant pressure to raise the temperature from a standard temperature to the given temperature. A computation of the temperature of the product gas, with an assumed percentage of the heat derived from the fuel remaining in the product gas, would be desirable. This might be done for some arbitrarily chosen fuel analysis, but for general purposes it is more convenient to make computations in an opposite direction. The enthalpy of that amount of product gas corresponding to 1 lb of the given fuel is calculated at a variety of excess-air levels and temperatures. Comparison of these values with the percentage of the heating value of the fuel that may be assumed to remain in the product gas gives an estimate of gas temperature.

Data needed for this computation are given in Table 2–4. Equations provided by Kelley [7] were converted to units of Btu per pound-mole and

Table 2–4   Enthalpies (Btu per pound mole over standard state[a])

Temperature T (°F)	CO_2	O_2	N_2	H_2O
1000	10,048	6,974	6,720	26,925
.1500	16,214	11,008	10,556	31,743
2000	22,719	15,191	14,520	36,903
2500	29,539	19,517	18,609	42,405

Enthalpy Equations

$$CO_2\ H = 10{,}570\left(\frac{T+460}{1000}\right) + 583.3\left(\frac{T+460}{1000}\right)^2 + \frac{667.4}{(T+460)/1000} - 7085$$

$$O_2\ H = 7{,}160\left(\frac{T+460}{1000}\right) + 278.8\left(\frac{T+460}{1000}\right)^2 + \frac{129.6}{(T+460)/1000} - 4163$$

$$N_2\ H = 6{,}830\left(\frac{T+460}{1000}\right) + 250.0\left(\frac{T+460}{1000}\right)^2 + \frac{38.9}{(T+460)/1000} - 3811$$

$$H_2O\ H = 7{,}300\left(\frac{T+460}{1000}\right) + 683.3\left(\frac{T+460}{1000}\right)^2 + 14{,}810$$

[a] Gas, except liquid water, at 1-atm pressure and 77°F.

degrees Fahrenheit. The chosen state, except liquid water, was gas at 77°F, the standard commonly used in determining the gross heating value of fuels.

The enthalpy of the gas at a given temperature and excess-air level may be determined from the relation

$$\left(\frac{\text{theoretical moles of product gas}}{\text{pounds of refuse}}\right) \left(\frac{\text{total moles of gas}}{\text{theoretical moles of gas}}\right)$$

[summation of (mole fraction of gas component)

$$(\text{Btu per mole of gas component})] = \frac{\text{Btu in product gas}}{\text{pounds of refuse}}.$$

The illustrative data needed for the calculation are found in Tables 2–1, 2–2, and 2–4. Note that mole fractions equal percentages divided by 100. For 50 percent excess air and 1500°F the result is

$$0.16044 \left(\frac{141.08}{100}\right)$$

$$[0.110(16,214) + 0.060(11,008) + 0.682(10,556) + 0.148(31,743)]$$
$$= 3246 \text{ Btu per pound of refuse.}$$

Data for two excess-air levels and for four temperatures are given in Table 2–5 for the particular refuse used in this illustration.

Table 2–5   Heat Content of Combustion Product Gases

Temperature (°F)	Heat Content of Gas from 1 lb of Refuse (Btu)	
	50% Excess Air	125% Excess Air
1000	2284	2973
1500	3246	4320
2000	4250	5721
2500	5294	7176

The gross heating value of the refuse was 5442 Btu/lb. This heat must be distributed between the product gas and ash residues or be lost through the incinerator walls. If the loss through walls and in ash residues is assumed to be 10 percent of the total, 4898 Btu per pound of refuse must remain in the gas. An interpolation between the data of Table 2–5 indicates that the gas temperature would be about 2300°F at 50 percent excess air or 1700°F at 125 percent excess air.

EMISSION OF AIR POLLUTANTS

Air pollutants from incineration may be classified roughly as those that result from incomplete combustion and those that, though they may be modified by combustion conditions, are not necessarily products of incomplete combustion. Among the latter are particulate matter, sulfur oxides, and nitrogen oxides.

Particulate Matter

Particulate matter includes fly ash, fragments of incompletely burned fuel, and particles of colloidal size related to smoke formation. The classification is empirical. It may include condensation products, such as moisture and tars, that were not necessarily present in particulate form when the gases were at a higher temperature and that may or may not be eliminated from the collected material by drying. The amount of particulate matter, even with a given refuse, varies over wide limits, depending on the completeness of combustion and the aerodynamic conditions governing the lifting of particulates from the fuel bed and their settling out from the combustion products.

Particulate emission is conventionally reported either as grains per standard cubic foot (gr/scf) or as pounds per 1000 lb of dry flue gas [8]. These quantities, in turn, may be calculated to such conditions as 50 percent excess air or to 13 percent carbon dioxide. The calculation to standard excess-air condition or a given amount of carbon dioxide in the flue gas is done to set limits that cannot be evaded by excessive dilution either by air or by steam. In this connection it should be recognized that a limitation based on grains per standard cubic foot requires definition as to standard pressure and temperature. The limits should be on a dry basis to prevent evasion through dilution with steam.

The particulate loading, as weight per unit volume at specified conditions, can be approximated directly from the observed particulate loading and the gas analysis. Calculation on the basis of pounds of particulate matter per 1000 lb of dry flue gas, however, requires knowledge of gas density, which in turn depends on the gas composition. Detailed calculations are illustrated first, followed by simple formulas that suffice when the corrected gas analyses and densities are not needed.

The calculation is illustrated for a flue gas with the composition given in Table 2–2 at 125 percent excess air. The composition, calculated to the dry basis, is listed in column 2 of Table 2–6. It is assumed that,

Table 2–6   Correction of Gas Composition to 50 Percent Excess Air, Dry Basis[a]

	Amount of Component (ft³)					Volume Percent
Component	In Product Gas	In Air Used	Difference	In Air To Be Subtracted	Corrected Com-position	Corrected Com-position
Carbon dioxide	8.59	0.03	8.56	0.01	8.58	12.91
Oxygen	11.73	21.15	−9.42	7.02	4.71	7.08
Nitrogen	79.68	79.68	0.00	26.48	53.20	80.01
Total	100.00	100.86		33.51	66.49	100.00

[a] Dry-air composition, in volume fractions: carbon dioxide, 0.0003; nitrogen, 0.7902; oxygen, 0.2905. Rare gases included with nitrogen.

through use of a condenser ahead of the gas meter and analyzer, the gas analysis was determined as given in Table 2–6, and the particulate loading was measured as 0.299 gr/ft³ (1 atm and 77°F) of dry gas. Note that sulfur dioxide has been included with carbon dioxide since these gases have the same oxygen requirement and appear together in an Orsat analysis. Percentages also have been recalculated to two decimal points for the purpose of comparison. It is assumed that all of the nitrogen in the gas was supplied as nitrogen in dry air. Column 3 of Table 2–6 lists the volume for each component of dry air required to supply the nitrogen. Column 4 lists the differences between column 2 and column 3. The percent of excess air is estimated from the ratio of the excess oxygen, 11.73 ft³, to that of oxygen used, 9.42 ft³. The estimated percent of excess air, $100(11.73/9.42) = 124.5$, differs from that given in Table 2–1 because of rounding errors and the assumption that all of the nitrogen came from the air. The amount of oxygen to be removed to reduce the excess to 50 percent is

$$11.73 \left(1.0 - \frac{50.0}{124.5}\right) = 7.02 \text{ ft}^3.$$

Column 5 lists the corresponding volumes for all components of dry air. Subtraction from column 2 gives the product-gas composition corrected to 50 percent excess air, at which the particulate loading is

$$\frac{0.299}{0.6649} = 0.450 \text{ gr/ft}^3 \text{ of dry gas.}$$

The fractional contraction is also given by the simple formula

$$F_{50} = 1 - \left(\frac{(O_2)}{20.95}\right)\left(\frac{E - 50}{E}\right).$$

This formula still contains the approximation that nitrogen from the fuel is ignored; it will not suffice if the corrected gas composition and density are needed.

The correction to 13 percent carbon dioxide is illustrated in Table 2–7. The calculation is the same as that for Table 2–6 except that the volume of air to be subtracted is given by the relation

$$\frac{(13.0 - 8.58)100}{13.0} = 34.00 \text{ ft}^3.$$

Allowance was made for carbon dioxide in the air by using the observed percentage of carbon dioxide less 0.01 as the actual carbon dioxide. The correct allowance can be estimated by a preliminary calculation without this correction. The particulate loading corrected to 13 percent carbon dioxide is $0.299/0.6600 = 0.453$ gr/ft^3. The same result is given by multiplying the grain loading by the ratio of carbon dioxide concentrations:

$$0.299 \left(\frac{13.0}{8.59}\right) = 0.453 \text{ gr/ft}^3.$$

The correction for carbon dioxide in the air is practically negligible. The simple correction also does not give the gas composition and density.

Table 2–7   Correction of Gas Composition to 13 Percent Carbon Dioxide, Dry Basis

| Component | Amount of Gas Component (ft³) | | | Volume Percent |
	In Product Gas	In Air To Be Subtracted	Corrected Composition	Corrected Composition
Carbon dioxide	8.59	0.01	8.58	13.00
Oxygen	11.73	7.12	4.61	6.99
Nitrogen	79.68	26.87	52.81	80.01
Total	100.00	34.00	66.00	100.00

Calculation of particulate loadings to a basis of pounds per 1000 lb of flue gas requires an estimate of the gas density. This may be obtained from the gas composition and the fact that the apparent molecular weight of mixed gases is, for ideal gases, a simple weighted average of the individual molecular weights. Thus, using the data of Table 2–6, we find that

$$0.1291(44.011) + 0.0708(32.000) + 0.8001(28.016) = 30.363$$

is the molecular weight of the dry gas.

The molecular volume, the volume containing 1 molecular weight of gas, depends on the pressure and temperature; Table 2–8 shows values at 1-atm

**Table 2–8   Molecular Volumes at
1 Atm and Various Temperatures**

Temperature (°F)	Molecular Volume (ft³ at 1 atm)
0	335.9
32	359.3
60	379.7
68	385.5
70	387.0
77	392.1

pressure and various temperatures. The calculation is according to the formula

$$\frac{1000(gr/ft^3)(ft^3/mole)}{(gr/lb)(average\ molecular\ weight)} = lb/1000\ lb$$

or for the present illustration (with the standard cubic foot taken at 1 atm and 77°F)

$$\frac{1000(0.450)(392.1)}{7000(30.363)} = 0.830\ lb\ per\ 1000\ lb\ of\ dry\ gas\ at\ 50\%\ excess\ air.$$

The data of Table 2–7 give 30.373 as the average molecular weight and 0.835 lb per 1000 lb of dry gas at 13 percent carbon dioxide.

The calculation of particulate loadings has been considered in some detail. This was done because the particulate loadings, calculated to standard conditions, are involved in regulations. It is important that there should be general understanding of the approximations and definitions that are used in these calculations.

Sulfur Oxides

For the purpose of estimating the total emission of sulfur dioxide and sulfur trioxide, retention of sulfur in ash residues and in products of incomplete combustion may be neglected as comparatively small, and the total content of these oxides may be estimated by mass balance from the sulfur content of the refuse. The calculation is included in Tables 2–1 and 2–2, although the sulfur concentration in the refuse, 0.12 percent by weight, is so low that the concentration in the flue gas is almost lost in numerical rounding errors. The concentration shown for sulfur dioxide in the product gas actually represents the total for sulfur dioxide and sulfur trioxide. The extra oxygen of the trioxide was neglected in the estimation of oxygen balances and air requirements. The oxidation of sulfur dioxide to sulfur trioxide is known to occur in flames [4] but otherwise requires catalysis. At equilibrium the ratio of sulfur trioxide to the total sulfur oxides depends on the temperature and the oxygen concentration of the flue gas. The relationship is shown in Figure 2–6. The sulfur trioxide in gases leaving flame is generally under 1 percent of the total sulfur oxides. The fraction should increase as the gases are cooled, but the reaction

$$SO_2 + \tfrac{1}{2}O_2 \rightarrow SO_3$$

becomes so slow that there is not time for the reaction to reach equilibrium. As a result the approach to equilibrium tends to "freeze" at temperatures on the order of 2000°F. About 5 to 10 percent of the sulfur oxides may appear as sulfur trioxide, although the equilibrium value is practically 100 percent at ambient conditions with excess air.

Nitrogen Oxides

Nitric oxide (NO) is generally formed in combustion processes. Nitric oxide in the atmosphere is slowly converted to nitrogen dioxide (NO_2). Nitrogen oxide emissions have been reported [5] on the order of 2 lb of equivalent nitrogen dioxide per ton of refuse. The relation to concentration may be shown by assuming that such an emission rate had been observed for the illustration used in Tables 2–1 and 2–2. In this illustration 650 moles of gas were produced per ton of refuse. The 2 lb of nitrogen dioxide corresponds to 0.043 mole. Hence 0.043/650 = 0.0007, or 700 ppm, of nitrogen oxides were found in the flue gas. Catalytic reactions favor decomposition of nitric oxide as the temperature falls. Nitric oxide formation

$$\tfrac{1}{2}N_2 + \tfrac{1}{2}O_2 \rightleftarrows NO$$

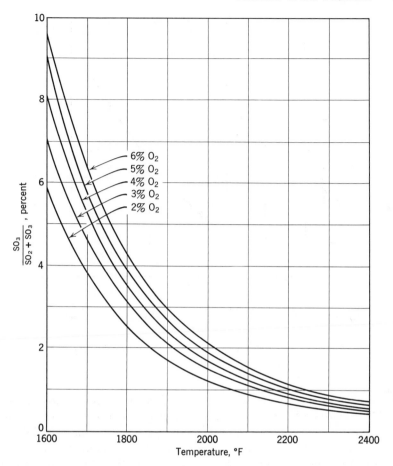

Figure 2–6 Equilibrium concentration of sulfur trioxide as percentage of total sulfur oxides.

tends to satisfy an equilibrium with a constant

$$K = \frac{NO}{(N_2)^{1/2}(O_2)^{1/2}}.$$

The dependence of this equilibrium constant on temperature is given in Figure 2–7. The apparent equilibrium constant for the illustration at hand is

$$\frac{0.07}{\sqrt{71.6}\,\sqrt{10.5}} = 0.0026,$$

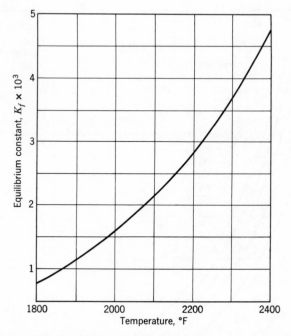

Figure 2–7   Equilibrium constant for nitric oxide formation.

which corresponds to an apparent equilibrium temperature of about 2150°F. Temperatures in local volumes of intense combustion were undoubtedly much higher. Correspondingly higher concentrations, approaching equilibrium values, were probably present in these high-temperature zones. Some combination of dilution by excess air and "freezing" of catalytic decomposition led to the apparent equilibrium temperature of 2150°F. A tendency appears for the nitric oxide decomposition to "freeze" at apparent equilibrium temperatures on the order of 2000°F. Order-of-magnitude estimates of nitrogen oxide concentrations can be made from the gas analyses by assuming $K = 0.0016$ and solving the equilibrium expression for the percent of nitric oxide. For a product gas with 6.0 percent oxygen and 68 percent nitrogen the nitric oxide concentration would be on the order of $(0.0016) \sqrt{6.0} \sqrt{68.0} = 0.04$ percent, or 400 ppm.

Smoke

Smoke is a suspension of particulate matter, liquid or solid, in the gaseous products of combustion. Visibility depends on the absorption and scattering of light, which in turn are more a function of the number of particles than of their weight. Hence the density of smoke is not a direct measure of the weight of particulate emissions. Particulate matter derived from fly ash plus fragments of partially burned fuel is not generally present in sufficient numbers of particles per unit volume to produce dense smoke. The production of dense smoke requires the condensation of liquids (commonly intermediate combustion products) into aerosol form, at least initially a dispersion of finely divided droplets. To understand smoke formation it is necessary to consider the processes involved in combustion that produce liquid phases and influence their aerosol condensation products.

All organic substances are thermally unstable. On heating to temperatures characteristic of each substance, pyrolysis, a combination of thermal cracking and condensation reactions, leads to a split into gaseous, liquid, and solid products. The gaseous and liquid products may be (a) promptly mixed with sufficient air at such temperatures that they are rapidly converted into products of complete combustion, (b) subjected to continuing pyrolysis, (c) condensed more completely to a liquid aerosol, or (d) subjected to some combination of these processes. Pyrolysis always involves a competition between thermal cracking, producing lighter and more stable products, and condensation reactions, producing heavier molecules so that the liquid droplets are converted to solid particles eventually approaching carbon. Depending on the details of heat and gas flow and the history with respect to temperature, oxygen content, fuel content, and pyrolytic processes in each stream of gas passing through the fuel bed and the incinerator chambers, any combination or stage of the above processes may be responsible for the smoke leaving the stack.

Within the broad definition given here the condensation of steam into a liquid aerosol produces a white-smoke plume. Lack of persistence, due to revaporization, distinguishes this plume from true smoke, which may be white to brown in color. This lightly colored smoke, distillation smoke, is formed from volatile matter that escapes the fuel bed in a gas stream that is neither ignited nor heated to sufficient temperature to cause continuing pyrolysis. Black smoke is formed when volatile matter is condensed to a liquid aerosol and then heated to such high temperatures in the absence of oxygen that the liquid droplets are converted by continuing pyrolysis into a solid, black, particulate form. Smoke may be eliminated by mixing of the gas with sufficient oxygen at a temperature high enough for complete combustion to follow rapidly. The necessary temperature depends on the degree of pyrolysis. Black smoke may require such high temperatures that, once it is formed, it is impracticable to destroy it by delayed combustion.

Other Products of Incomplete Combustion

Smoke is often described as a product of incomplete combustion. Although this is a proper attribute, it is far from a complete description, as indicated in the above discussion. The thermal cracking and condensation reactions—either incidental to, or involved in, smoke formation—are also the source of a large variety of organic compounds. Any of these compounds that are either not burned or incompletely burned and are not condensed into smoke may escape as gaseous organic compounds. Without attempting a definitive or comprehensive listing, aldehydes (specifically formaldehyde), amines, organic acids, and hydrocarbons have been found; many others must be present, although mostly in insignificant amounts.

Carbon monoxide can also be classed as a product of incomplete combustion. It appears in considerable concentration when the air supply is below the theoretical requirement. It also appears, together with smoke, when fuel-rich streams issue from the fuel bed and are not properly mixed with the air necessary for complete combustion. Since the same conditions are involved in the formation of smoke and other products of incomplete combustion, some correlation with carbon monoxide formation should be expected. It has been found in fact that the percentages of carbon monoxide are relatively low for large incinerators with capacities ranging from 50 to 250 tons/day and that the emissions of carbon monoxide and hydrocarbons are significantly correlated [5].

The production of polynuclear hydrocarbons is probably related to the condensation reactions in the liquid aerosols that eventually lead to black smoke. In the study of a large pulverized-coal-fired furnace it was found in one test that the emission of polynuclear hydrocarbons, particularly pyrene, was about fivefold that for other tests [3]. This test involved maldistribution of air and coal, with the appearance of soot deposits on probes inserted at the furnace outlet. It was concluded that the polynuclear hydrocarbons were intermediates in the formation of black smoke [10]. As such, they are produced in the condensation reactions associated with pyrolysis; since they are relatively volatile, they may be vaporized and remain as gases until condensed in the sampling apparatus.

The formation of the products of incomplete combustion, typified by smoke, depends on details of heat and mass flow, and the aerodynamics of mixing processes. These processes involve so many variables and boundary conditions that quantitative correlation between emission rates and fuels, equipment design, and operational factors should not be expected. Nevertheless, the means for their control are well known.

Since the gaseous products of incomplete combustion generally are produced in processes associated with smoke, they may be expected to appear in amounts that rise and fall with smoke, and to respond to common control. As discussed in the section on management of fuel and air every stream of gas leaving the fuel bed must be mixed promptly with sufficient air at a temperature that is high enough to completely and rapidly burn the volatile matter present. It is in this area, particularly, that combustion becomes an engineering art rather than a science.

REFERENCES

[1] Bailie, R. C., D. M. Donner, and A. F. Galli, "Potential Advantages of Incineration in Fluidized Beds," *Proc. 1968 Nat. Incinerator Conf.*, New York, May 5–8, 1968, pp. 12–17.

[2] Carman, E. P., E. G. Graf, and R. C. Corey, "Combustion of Fuels in Thin Beds," *U.S. Bureau of Mines Bull.*, 563 (1957).

[3] Cuffe, S. T. R. W. Gerstle, A. A. Orning, and C. H. Schwartz, "Air Pollutant Emissions from Coal-Fired Power Plants," Report No. 1, *J. APCA*, **14** (9), 353–362 (September 1964).

[4] Dooley, A., and G. Whittingham, "Oxidation of Sulfur Dioxide in Gas Flames," *Trans. Faraday Soc.*, **42**, 354–366 (1946).

[5] Hangebrauck, R. P., D. J. von Lehmden, and J. E. Meeker, "Emissions of Polynuclear Hydrocarbons and Other Pollutants from Heat Generation and Incineration Processes," *J. APCA*, **14**, 267–278 (July 1964).

[6] Kaiser, Elmer R., "Chemical Analysis of Refuse Components," *ASME Preprint 65-WA/PID-9*, presented at ASME Winter Annual Meeting, November 7–11, 1965.

[7] Kelley, K. K., "Contributions to the Data on Theoretical Metallurgy," *U.S. Bureau of Mines Bull.*, 584 (1960).

[8] New York City Air Pollution Control Code, Section 9.09a, effective October 1, 1964.

[9] Nicholls, P., "Underfeed Combustion, Effect of Preheat, and Distribution of Ash in Fuel Beds," *U.S. Bureau of Mines Bull.*, 378 (1934).

[10] Orning, A. A., C. H. Schwartz, and J. F. Smith, "Minor Products of Combustion in Large Coal-Fired Steam Generators," *ASME Preprint 64-WA/FU-2*, presented at ASME Winter Annual Meeting, November 29–December 4, 1964.

[11] Weintraub, M., A. A. Orning, and C. H. Schwartz, "Experimental Studies of Incineration in a Cylindrical Combustion Chamber," *U.S. Bureau of Mines Rept. Inv.*, 6908 (1967).

3

PRINCIPLES OF CLEANING COMBUSTION PRODUCTS

Joseph H. Field

In the incineration of waste material problems often arise because of the emission of particulate matter and foul odors that result from the improper design or operation of incineration equipment. Investigations of these factors have been made to minimize pollution [19, 29, 34]. It has become common practice to employ collecting devices to decrease the emission of particulate matter and to reduce the emission of noxious gases by designing and operating the incinerator for maximum combustion efficiency. The fundamental principles involved in control devices, including basic particle dynamics, and the principles utilized in odor control are covered in this chapter.

Because of the variable nature of the materials charged to incinerators and the changing mode of operation, the rate of particulate emission may vary widely; for example, values of emission from municipal incinerators are reported at 3 to 5 lb per ton [42] to 10 to 25 lb per ton of refuse burned [9, 23]. There is general agreement, however, that the

Project Coordinator, Process Engineering, Pittsburgh Coal Research Center, Bureau of Mines, U.S. Department of the Interior, Pittsburgh, Pennsylvania.

size of the particles is quite small. Flood [9] has reported that the particles emitted from incinerators range from less than 5 microns (0.0002 in.) to about 120 microns (0.005 in.) in diameter, with about one-third of the total less than 10 microns in diameter. These particles fall generally in the range of fine dust, as shown in the classification chart of Figure 3–1, prepared by Lapple [22].

Removal of particles with diameters smaller than 50 microns is difficult and requires efficient collecting devices. Basic understanding of particle dynamics and the physical principles applied in the various types of control devices is necessary in evaluating equipment for removing particulate matter emitted in a specific incinerator. Devices for removing particulate matter are based on gravity settling of particles, inertia or momentum, filtration, electrostatic precipitation, and agglomeration of fine particles by sonic or mechanical means to facilitate removal by increasing the particle size. Both wet and dry media can be employed in most of these devices.

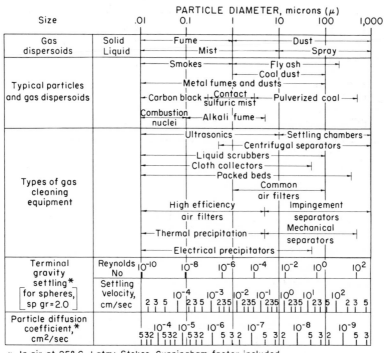

Figure 3–1 Particle-classification chart.

For controlling or eliminating objectionable odors emanating from incineration, under circumstances where they cannot otherwise be controlled, dispersal, adsorption, secondary combustion (or afterburning) with or without catalysts, odor modification or masking, and wet scrubbing are employed [45]. Important aspects of these methods of odor control are discussed later. Some idea of the problems faced in odor control can be realized from the analyses (Table 3–1) of gases from a typical municipal incinerator and a single-chamber, flue-fed incinerator

Table 3–1   Emissions from Municipal and Apartment-Building Flue-Fed Incinerators[a]

	Municipal Incinerator[c]	Apartment Flue-Fed Incinerator[b]	
		Standard	With Two-Pipe Overfire Jets
Aldehydes	1.1	4.6	2.5
Acids	0.6	22.4	10.4
Other organics (esters, phenols, etc.)	1.4	21.6	14.0
Ammonia	0.3	0.4	0.5
Nitrogen oxides	2.1	0.1	0.2
Sulfur oxides	1.9	0.5	0.2
Total noxious gas	7.4	49.6	27.8

[a] Units are pounds per ton of refuse.
[b] Data from [19].
[c] Data from [14].

commonly used in apartment buildings to burn refuse. The results of a test with overfire-air jets to decrease emission from the flue-fed incinerator are included in the table.

PARTICLE DYNAMICS

Particles carried in a gas stream are subject to many forces and actions. They tend to settle due to gravity, collide with other particles and possibly agglomerate, and may collide and impinge on obstacles such as walls, baffles, and liquid droplets. By utilizing these forces and abetting them with centrifugal, electrostatic, and sonic actions, separation of the particles from the gas stream is achieved. The forces acting on entrained particles are covered in this section on particle dynamics in order to elucidate the principles utilized in separation devices. Special emphasis is given to the dynamics of particles 10-micron diameter or smaller because of their prevalence in incinerator emissions.

Free Fall of Particles

The fundamentals of the free fall of particles in a gas can be handled most readily by analyzing the dynamics of a single spherical particle. When there is relative motion between a fluid and a particle immersed in the fluid, such as with free-falling solids, drag forces are exerted on the particle. These forces are caused by skin friction and the differential pressure upstream and downstream of the particle that results from the change in the streamlines of the gas.

A free-falling particle accelerates until the frictional drag of the fluid surrounding the particle balances the gravitational force. From this point the particle continues to fall at a constant velocity, called the terminal velocity. The terminal velocity of a free-falling sphere can be calculated from the following:

$$f = \frac{\pi}{8} C_D \rho_f d_p^2 u_t^2 = V_p(\rho_p - \rho_f)g$$

or

$$u_t^2 = \frac{V_p(\rho_p - \rho_f)g}{\pi/8 C_D \rho_f d_p^2}, \tag{3-1}$$

where f = drag force,
C_D = drag coefficient,
ρ_f = fluid density,
d_p = particle diameter,
u_t = terminal velocity of particle,
V_p = particle volume,
ρ_p = particle density,
g = gravitational constant.

Since the volume of a sphere, $V_p = (\pi d_p^3/6)$,

$$u_t^2 = \frac{4}{3} \frac{d_p(\rho_p - \rho_f)g}{\rho_f C_D}. \tag{3-2}$$

The drag coefficient C_D is a function principally of the Reynolds number, which in turn is related to the diameter of the falling particle. The three ranges of Reynolds numbers considered in determining drag coefficients are clearly illustrated in a publication by Leniger [24] in the following manner:

Reynolds Number	10^{-4}	10^0	10^3	10^5 or larger
Range	Stokes' law applies	Transition range	Newton's law applies	
C_D	$\dfrac{24}{N_{\mathrm{Re}}}$	$f(N_{\mathrm{Re}})$	0.43	
u_t	$\dfrac{g\Delta\rho\, d_p{}^2}{18\mu_f}$	$\left(\dfrac{4\Delta\rho}{3C_D\rho_f}d_pg\right)^{\frac{1}{2}}$	$1.74\left(\dfrac{\Delta\rho}{\rho_f}d_pg\right)^{\frac{1}{2}}$	
Controlling resistance	Frictional	Frictional plus shape	Shape	

The Reynolds numbers below 1 correspond to particle diameters of 100 microns or less; the maximum Reynolds number in the transition range of 1000 corresponds to a particle diameter of about 2.5 mm; higher Reynolds numbers, to particles larger than 2.5 mm. For small particles

$$C_D = \frac{24}{N_{\mathrm{Re}}} = \frac{24\mu_f}{d_p u \rho_f}$$

so that by substituting for C_D in the previous equation we obtain

$$u_t = \frac{g d_p{}^2(\rho_p - \rho_f)}{18\mu_f}, \tag{3-3}$$

which, as shown in the table, is the Stokes'-law expression for the terminal velocity of a particle. The above relationship is based on a single sphere falling in a limitless gas expanse unhindered by other particles or boundary surfaces. Use of this equation can be exemplified as follows: Given a 100-micron (3.27×10^{-4} ft) spherical particle with a true specific gravity of 85 lb/ft³ and falling in air at 70°F, the terminal velocity is

$$u_t = \frac{(32.2)(3.27 \times 10^{-4})^2(85 - 0.08)}{18(0.018 \times 0.672 \times 10^{-3})} = 1.41 \text{ ft/sec.}$$

Drag coefficients have been determined for spheres over a wide range of Reynolds numbers (up to 3 million), as well as for other shapes such as horizontal and vertical disks, cylinders, cubes, tetrahedrons, and octahedrons. A plot of the drag coefficient with Reynolds number for spheres, disks, and cylinders is shown in Figure 3-2. For nearly spherical particles the terminal velocity is calculated by using the diameter of the equivalent sphere.

Very small particles, whose diameter approaches the mean free path of the gas molecules, fall faster than would be predicted from Stokes' law. This occurs because the particles slip between the gas molecules.

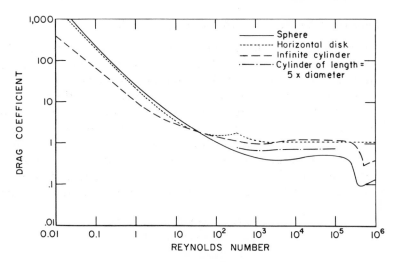

Figure 3–2 Drag coefficients as a function of Reynolds number.

The Cunningham [6] correction factor is applied to Stokes' law for such particles and results in the following equation for calculating terminal velocity:

$$u_t = \left(1 + \frac{K\lambda}{d_p}\right) \frac{g d_p{}^2 (\rho_p - \rho_f)}{18\mu_f}, \tag{3-4}$$

where K is a constant ranging from 1.25 to 2.3 for different gases and particle sizes, and λ is the mean free path of the fluid molecules.

As shown in Table 3–2, correction for small particles in air is considerable, but the correction approaches unity with increasing size.

Table 3–2   Cunningham Correction Factors[a,b]

Particle Diameter (microns)	Cunningham Correction $\left(1 + \dfrac{K\lambda}{d_p}\right)$
0.01	22.35
0.1	2.87
1.0	1.16
10.0	1.016
20.0	1.008

[a] Data from [54].
[b] Mean free path of the fluid molecules, λ, is 6.53×10^{-6} cm.

Effect of Brownian Motion on the Settling of Particles

Another factor that affects the settling of small particles is their random Brownian motion resulting from bombardment by the gas molecules. The Brownian effect is significant for particles of 1-micron size and smaller, and becomes increasingly important as the size of the particle decreases. A measure of the effect of Brownian motion is the displacement ratio, which is defined as the ratio of the random movement of a falling particle due to Brownian motion as compared with the distance it falls by gravity in the same time. Variation of the displacement ratio with size of the particles for particles falling in air is shown in Table 3–3. A general formula expressing the displacement of particles

Table 3–3   Displacement Ratio for Various Sizes of Particles in Air[a]

Particle Size (microns)	Displacement Ratio
0.1	17
0.5	0.45
1.0	0.09
5.0	0.002

[a] Data from [27].

due to Brownian forces based on the theoretical derivations by Einstein [8] and Von Smoluchowski [47] is

$$x = \left[\left(\frac{2RT}{N} \right) \left(\frac{t}{3\pi\mu_f d_p} \right) \left(1 + \frac{K\lambda}{d_p} \right) \right]^{\frac{1}{2}}, \qquad (3\text{–}5)$$

where R = the gas constant,
          T = the absolute temperature,
          N = Avogadro's number,
          t = time,
          μ_f = gas viscosity,
          d_p = particle diameter,
$(1 + K\lambda/d_p)$ = the Cunningham correction factor [6].

For settling of spherical particles in air at 20°C the equation becomes [25]

$$x = 6.8 \left[\frac{t}{d_p} \left(1 + \frac{K\lambda}{d_p} \right) \right]^{\frac{1}{2}}, \qquad (3\text{–}6)$$

where x, the average displacement and d_p, the particle diameter, are in microns and t, the time, is in seconds.

Thus, for example, a particle of 0.1-micron diameter settling in air is displaced randomly about 36 microns in 1 sec, as calculated by $x = 6.8[(1/0.1)(2.87)]^{1/2} = 36.4$. Since the displacement ratio of a particle of this size is about 17, the particle would settle only about 2 microns in this same time, illustrating the significance of Brownian motion for small particles.

Agglomeration of Particles

Agglomeration of micron and submicron particles into a cluster as a result of collision and adherence often can facilitate removal from a gas stream. Collision of small particles usually is brought about either as a result of Brownian motion or a velocity gradient. In the latter case the particles tend to collide because they are moving at varying velocities in adjacent streamlines. A third means of causing agglomeration is by use of sonic or ultrasonic energy. The factors that affect agglomeration of small particles are particle size, particle nature surface, adsorbed gases on the surface, state of dispersion, humidity, and possibly the temperature and viscosity of the gas medium. As an approach to understanding the adhesion of particles, Bradley [2] and Hamaker [11] derived an expression of the adhesive force between two bodies based on the London-Van der Waals forces of attraction as follows:

$$f = \frac{\pi^2 q_0^2 \lambda'}{12X^2} \frac{d_1 d_2}{d_1 + d_2}, \qquad (3\text{-}7)$$

where f = force of attraction in dynes,

q_0 = number of atoms per cubic centimeter of the substance,

λ' = London-Van der Waals constant,

d_1, d_2 = diameters of the particles in centimeters,

X = distance between particles in centimeters.

Hamaker reported the value of $\pi^2 q_0^2 \lambda$ as 10^{-12}. Corn [5], however, after a thorough study of the literature on particle cohesion and adhesion reported that the above equation cannot be used for predicting quantitatively the adhesion forces of small particles in air because adsorbed gases prevent close approach of solid surfaces.

Brownian coagulation of smokes was studied intensively by Whytlaw-Gray [51] and his co-workers, who derived the expression

$$\frac{1}{n} - \frac{1}{n_0} = K_B t, \qquad (3\text{-}8)$$

where n = the number of particles per cubic centimeter at time t,

n_0 = the original concentration,

K_B = the agglomeration constant.

Differentiating the equation gives an equation relating the rate of agglomeration to the square of the number of particles present,

$$-\frac{dn}{dt} = K_B n^2. \tag{3-9}$$

The rate of agglomeration resulting from a velocity gradient, sometimes called turbulent coagulation, also is proportional to the square of the concentration of particles and is expressed by

$$-\frac{dn}{dt} = K_S n^2. \tag{3-10}$$

For smoke experimental values of K_S have been found to be several times those of K_B for Brownian coagulation [44].

Adhesion of Particles to Solid Surfaces

Adhesion of small particles to solid surfaces is another aspect of particle dynamics and is important in separation techniques such as filtration, impaction, and electroprecipitation. Particle size and shape, particle and surface material, surface roughness, temperature and humidity of the ambient gas, time of particle-surface contact, and static electricity are known to influence adhesion. Corn [5] investigated the effect of several of these variables on the force of adhesion of quartz and Pyrex-glass particles with diameters from 20 to 90 microns on a Pyrex flat or a glass microscope slide. His findings can be summarized as follows:

1. *Effect of particle size.* The force of adhesion is proportional to particle size; for example, for quartz particles adhering to an optical flat $f = 0.012d_p$, where f is the force of adhesion in dynes and d_p is the particle diameter in microns.

2. *Effect of relative humidity.* Adhesion increases with increasing relative humidity, as shown in Figure 3-3. The increased adhesive force is believed to result from the increase in the thickness of the moisture film at the higher humidity.

3. *Effect of surface roughness.* As might be anticipated, adhesion decreases with increased surface roughness. Corn's experiments for adhesion of quartz particles to Pyrex plates were correlated by the expression

$$S = 8.8 \times 10^3 e^{-0.053 A}, \tag{3-11}$$

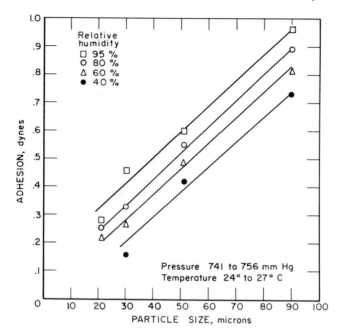

Figure 3–3   Adhesive force of quartz particles to Pyrex flat.

where S is the average surface roughness in angstroms and A, the percent of adhesion.

4. *Effect of time of contact.* The force of adhesion was not significantly altered by the time of contact if the contact time was at least 5 min.

Settling with the Aid of Centrifugal Force

A particle entrained in a gas flowing in a free vortex path has horizontal components in a tangential direction u_{tan} due to momentum and in a radial direction u_r due to centrifugal and centripetal forces, as well as a gravitational vertical component u_g. These components are shown in Figure 3–4. Centripetal force is provided by the drag resistance of the gas opposing the radial outward centrifugal force. At a given velocity in the vortex path the movement in the horizontal plane may be either radially inward or outward, or the particle may remain in a fixed orbit, depending on the magnitude of the opposing radial forces.

Particles whose mass and momentum are sufficiently high move toward the outer perimeter and travel outward until they reach a stable orbit or are restrained by a reactive force such as the wall of a containing vessel.

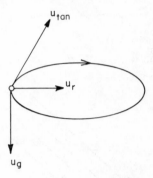

Figure 3–4 Velocity components of particle flowing in free vortex.

The centrifugal force on the particle is calculated from

$$f = \frac{V_p u_{\tan}^2 (\rho_p - \rho_f)}{r},$$ (3–12)

where V_p is the particle volume and r is the radius of the path. For a sphere of diameter d_p

$$f = \frac{\pi d_p{}^3 u_{\tan}^2 (\rho_p - \rho_f)}{6r}.$$ (3–12a)

As shown by the equation, the centrifugal force increases with increasing particle diameter and density, and increasing tangential velocity; it decreases with increasing radial path.

An expression for the centrifugal-settling velocity of a particle can be obtained by equating the centrifugal force to the resistive drag. This represents the condition where the two forces are in balance, and the particle is following an equilibrium orbit. Assuming Stokes' law, the resistive-drag force $f = 3\pi\mu_f u_{ct} d_p$, where μ_f is the dynamic viscosity of the fluid and u_{ct} is the centrifugal-settling velocity; then

$$u_{ct} = \frac{d_p{}^2 (\rho_p - \rho_f)}{18\mu_f} \frac{u_{\tan}^2}{r}.$$ (3–13)

The centrifugal-settling velocity can be compared to the terminal settling velocity in Stokes' region (3–3) by the following:

$$u_{ct} = \frac{u_{\tan}^2 u_t}{rg}.$$ (3–14)

Impingement

Removal of dust particles by impingement is utilized principally in filtration devices, baffled chambers, and wet scrubbers. Separation of the particles occurs because the inertia of the particles contained in a flowing gas is greater than that of the gas. Thus, as shown in Figure 3–5, a dust-laden gas moving toward an object is deflected around the object, whereas a portion of the particles will collide with the object. The solid lines represent the gas streamlines, and the dashed lines the particle trajectories. In the case illustrated, for a gas moving at velocity u toward a cylinder of diameter D, the particles contained in the portion of gas b, defined by streamlines A and B, will impinge.

The effectiveness of impingement, designated as collision cross section or efficiency of collision η, is defined as the fraction of particles carried by the gas that impinge on the object; for a cylindrical target it is

$$\eta = \frac{b}{D},\tag{3-15}$$

for a spherical target,

$$\eta = \frac{b^2}{D^2}.\tag{3-16}$$

Herne [12] has thoroughly reviewed the mathematical approaches to calculating the aerodynamic capture of particles by impingement on spheres. The following assumptions are made in the mathematical treatment:

1. Only two particles are considered.
2. One particle is much larger than the other, and the larger particle is spherical.
3. The fluid flow pattern is characterized by the flow around the larger particle.

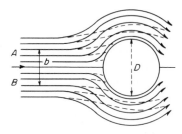

Figure 3–5 Streamlines on impingement. Solid lines—per streamlines; dashed lines—particle trajectories.

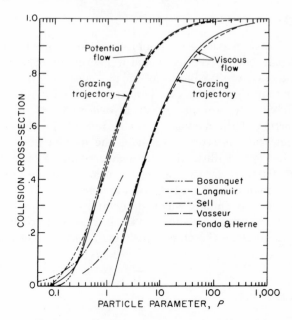

Figure 3–6 Variation of collision cross section with particle parameter.

4. The fluid forces on the smaller particle follow Stokes' law; that is, forces are linearly proportioned to the vector of the relative velocity of the particle in the fluid.

Using these assumptions, Sell [37] showed that impingement efficiency was a function of the dimensionless group mu/kd'_p, called the particle parameter P, where m is the mass of the particle, u is the gas velocity, k is the Stokes'-law coefficient, and d'_p is the diameter of the larger particle. For particles obeying Stokes' law $k = 3\pi\mu_f d_p$, and P becomes

$$P = \frac{\rho_p d_p{}^2 u}{18\pi\mu_f D}. \tag{3-17}$$

Figure 3–6 shows the correlation, as given by Herne [12] with data from several sources, between the collision cross section and the particle parameter for viscous and turbulent flow. The efficiency of collision and hence of particle removal by impingement is a function of the particle parameter, the Reynolds number (based on the diameter of the target), and the ratio of the diameters of the particle and the target, d_p/D. Since the collision cross section decreases with decreasing particle parameter, as shown in Figure 3–6, removal of particles by impingement becomes more difficult as the particle diameter decreases. Generally removal by this mechanism is negligible for particles of 0.5-micron diameter or smaller.

Electrostatic Forces

Particles entrained in a gas stream become charged when passed between two electrodes across which a high d-c voltage potential is impressed. The attraction of the charged particle to the grounded electrode (positive polarity) of the electrostatic precipitator results in removal of the particle. This is shown schematically in Figure 3–7. When the high negative voltage is applied to the discharge electrode negative gas ions are formed near the electrode surface. These ions become attached to the dust particles, giving them a negative charge. Underwood [46] points out that for particles of about 1-micron diameter or larger charging of the particle results mainly from collision with the ions in the field between the electrodes, whereas particles smaller than about 0.2 micron are charged principally by ion diffusion. For intermediate sizes of 0.2 to 1 micron White [49] showed that charging occurs by both mechanisms. A thorough description of the theory of particle collection by electrostatic precipitation is given in White's book [48].

A particle diameter of d_p subjected to a field strength of E_0 can attain a limiting charge Q_0 when charged by bombardment. Thus

$$Q_0 = n'e = \frac{pE_0d_p{}^2}{4},\qquad(3\text{–}18)$$

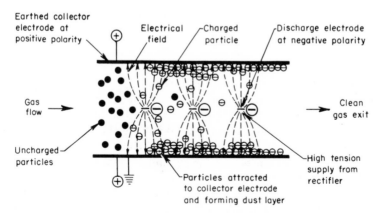

Figure 3–7 Principle of electrostatic precipitation of particles.

where n' = number of electrons acquired by the particle,

e = charge per electron,

p = factor $3E/(\epsilon + 2)$, where ϵ is the dielectric constant of the particle.

Under the influence of a collecting field of E_P (esu) the force on a particle of charge Q moving toward the collector is

$$f = QE = n'eE_P. \tag{3-19}$$

Assuming that Stokes' law is valid, the force on this particle is also $f = 3\pi\mu_f d_p u_t$, where u_t is the terminal velocity.

By equating these expressions and solving for u_t, the following is obtained:

$$u_t = \frac{n'eE_P}{3\pi\mu_f d_p}. \tag{3-20}$$

Then, by substituting for $n'e$, we obtain

$$u_t = \frac{pE_0E_P d_p}{12\pi\mu_f}. \tag{3-21}$$

In most cases the charging field E_0 and the collecting (or precipitating) field E_P are the same. For the smaller particles charged by ion diffusion Underwood [46] gives an expression for calculating the terminal velocity as follows:

$$u_t = \frac{eE10^6}{3\pi n'}. \tag{3-22}$$

DEVICES FOR SEPARATING PARTICULATE MATTER

Gravity-Settling Chamber

The simplest and cheapest device for removing dusts from gases is the settling chamber. As shown in Figure 3–8, this may be merely a large chamber or an enlargement in a duct that permits the gas velocity to be decreased sufficiently for particles to settle by gravity before the gas leaves the enlarged cross section. Obviously this restricts application to relatively large particles, and gravity-settling chambers are of little value for incinerator emissions; for example, a 100-micron-size spherical particle of unit density falls by gravity about 1 ft/sec in still air or when entrained in a stream of air moving horizontally without turbulence. Thus all particles of 100 microns or larger contained in a gas flowing with a horizontal gas velocity of 10 ft/sec in a 1-ft high duct would be

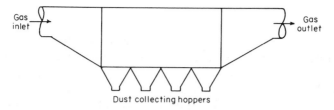

Dust collecting hoppers

Figure 3–8 Simple gravity-settling device.

deposited within 10 ft after the entry. By contrast, particles of 5-micron size fall at a rate of only 0.03 in./sec.

The efficiency of a gravity-settling chamber is denoted by

$$\eta = \frac{h}{H} = \frac{u_t A}{q},$$ (3-23)

where η = the efficiency for particles of a given diameter d,
  h = the height that particles of diameter d will settle in the time of travel through the chamber,
  H = the height of the chamber,
  u_t = the terminal velocity of particles of diameter d,
  A = the chamber cross section,
  q = the volumetric rate of flow.

Expressions for calculating the terminal velocity of particles have been already given in this chapter. For small particles where Stokes' law is valid, u_t is obtained from (3-3). By substituting the given expression for u_t and setting $\eta = 1$, the minimum particle diameter that theoretically will settle in a given cross-sectional area can be computed from the equation

$$d_p = \left[\frac{18\mu_f q}{g A (\rho_p - \rho_f)} \right]^{1/2}.$$ (3-24)

Because of turbulence in a moving gas stream and differences in the aerodynamic characteristics of particles, the speed of fall of a given-size particle can vary considerably from the value calculated by using Stokes' law. Therefore a settling chamber cannot collect all particles above the size for which it is nominally designed. However, the collection efficiency is still of value for comparing the performance of gravity collectors. Typical collection efficiencies of several settling chambers in industrial practice are given in a review by Jackson [16] for a range of particle sizes and densities.

Cyclone

The cyclone is one of the commonly used devices for separating solids from gases because of its relatively low capital and operating costs. Generally a cyclone consists of a vertical cylinder provided with a tangential entry positioned above a conical reducing section that leads to a hopper for the solids. Gas and entrained solids flow in a vortex path, with the larger particles separated by the centrifugal forces as already described. The clean gas leaves the cyclone at the top through the centrally located outlet that projects into the cylindrical section. Dimensions of a typical cyclone are shown in Figure 3–9. A multicyclone collector consisting of several cyclone tubes usually 9 to 10 in. in diameter, mounted on a common tube sheet, has been frequently used for collection of fly ash for incinerators [23].

The principal factors considered in the design of cyclones are the effectiveness of separation, the capacity, and pressure drop. The most commonly used criterion for evaluating cyclone performance is based

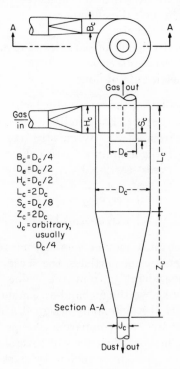

$$B_c = D_c/4$$
$$D_e = D_c/2$$
$$H_c = D_c/2$$
$$L_c = 2 D_c$$
$$S_c = D_c/8$$
$$Z_c = 2 D_c$$
$$J_c = \text{arbitrary, usually } D_c/4$$

Figure 3–9 Typical cyclone and dimensions.

on the empirical equation developed many years ago by Rosin, Rammler, and Intelmann [36]. Their derivation assumes that the gas moves through the cyclones as a rigid spiral maintaining a constant velocity equal to that at the cyclone inlet. This expresses the smallest particle diameter d_p that can be removed as

$$d_p = \left[\frac{9B_c\mu_f}{\pi N_s u_c(\rho_p - \rho_f)} \right]^{\frac{1}{2}}, \qquad (3\text{-}25)$$

where B_c = the diameter of the cyclone inlet,
   N_s = the number of spirals made by the gas in the cyclone (usually between 3 to 5),
   u_c = average gas velocity at the inlet.

Because the actual flow pattern in a cyclone can be considerably different from that used to derive the equation, values calculated from the equation can be considered estimates in many instances. The following sample problem illustrates application of the formula for sizing cyclones. The amount of dust in air is 4 gr/ft³, with 70 percent of the dust particles being 10 microns or larger. Solids density is 90 lb/ft³. The air temperature is 200°F, and the flow is 1950 ft³/min. Using the cyclone configuration of Figure 3-9 with $N_s = 5$, what cyclone diameter is required to remove 10-micron (3.27×10^{-5} ft) and larger particles, and how many parallel streams are needed? Use an inlet velocity of 40 ft/sec.

$$B_c = \frac{(3.27 \times 10^{-5})^2 \pi (5)(40)(90 - 0.06)}{9(1.48 \times 10^{-5})}$$

$$= 0.456\text{-ft, or } 5.48\text{-in. width of inlet duct.}$$

Then $D_c = 4(5.48) = 21.92$-in. cyclone diameter and $H_c = 2(5.48) = 10.96$ in. of height of inlet duct.

Since the inlet-duct cross section is 0.416 ft² and the inlet gas velocity is 40 ft/sec, the throughput per cyclone is 995 ft³/min. Thus two cyclones in parallel are required.

In addition to the 10-micron and larger particles, a portion of the smaller particles is removed in the cyclone. The amount of removal of the smaller particles can be estimated from a plot of the fractional weight-collection efficiency with the ratio of the particle size of the particle in question to the cut size used as a basis for design [43].

The cyclone diameter is the most important design factor affecting the efficiency of removal. For a given pressure differential across the cyclone the efficiency increases as the diameter decreases, because as $u_{tan}r$ is constant, the tangential velocity and thus the centrifugal force increases with decreased radius of rotation. For this reason the trend is to build cyclones of small diameter. This often necessitates use of multiple cyclones in parallel to maintain a given volumetric flow of gas at a limited pressure differential.

The pressure drop through a cyclone results from entrance and exit losses, friction along the wall, and loss of rotational kinetic energy in the gas. Generally the pressure drop is expressed in terms of the velocity head based on the inlet area [18]. The inlet velocity head Δh in inches of water can be computed from the equation of Shepherd and Lapple [39]:

$$\Delta h = 0.003\rho_f u_c^2, \tag{3-26}$$

where ρ_f is the gas density in pounds per cubic foot and u_c is the average inlet gas velocity in feet per second. The friction loss along the wall generally is 1 to 20 times the inlet velocity head and was found by Shepherd and Lapple [39] to be computed by

$$F_{cv} = \frac{K_c B_c H_c}{D_e^2}, \tag{3-27}$$

where F_{cv} is expressed as the number of inlet velocity heads and B_c, H_c, and D_e are values from the cyclone geometry (see Figure 3-9). The constant K_c equals 16.0 for the normal cyclone arrangement in which the rectangular inlet terminates at the outer element of the cylindrical cyclone body. Thus for the dimensions in Figure 3-9 the friction loss is 8 inlet velocity heads. For the example cited for computing efficiency the inlet velocity head is 0.288 in. of water and the friction loss is 2.30 in. of water. The constant K_c is decreased to 7.5 by extending the inner side of the inlet past the cylindrical wall into the annular space halfway to the wall.

Filters

Filtration can be applied to the removal of large particles, but it is particularly advantageous for separating small particles of submicron size. A filter consists of many small, cylindrical fibers over which the gas passes in streamline flow, and the entrained dust particles collide with the fibers. The inertia effect discussed in the description of impingement is important in the capture of larger particles. For particles of 1 micron and smaller Brownian motion causes deviation from the streamline path, increasing the probability of collision with the fibers.

Two common types of filters are fiber and fabric filters. The fiber filter consists of a loose or fabricated mass of fiber, whereas the fabric type is woven. Pilpel [31] describes several types of filters and shows their industrial application. Fibrous filters generally are used for low dust concentrations. As with cyclones, the principal criteria in specifying filters are the efficiency of removal and pressure drop. These two are often closely related in the case of filters because the pores of the filter become clogged with deposited dust, resulting in increased efficiency and pressure drop. Often formation of a precoat of captured particles is a requisite for efficient removal. In the usual case, after the pressure drop reaches a level at which power consumption becomes excessive, a cleaning cycle is started by using mechanical agitation, back flow, or shock waves. If the filter is not cleaned, a breakthrough of particles may occur.

A third item necessary for the design of a filter is the allowable gas throughput to compute the required filter area. For fibrous filters the gas velocity is usually specified, varying between 30 to 60 ft/min for the highly efficient types that are required for removing small particles and 300 to 500 ft/min for coarser filters. The gas throughput for fabric filters is commonly expressed by the filter ratio, which is defined as the ratio of the gas volume to gross filter area [35]. Filter ratios of 1 to 6 ft^3/(min)(ft^2 of fabric area) are common, the lower range being used for finer dusts.

The basic equation for computing the efficiency of a filter is

$$\ln \frac{N_1}{N_2} = \frac{4H\alpha\eta}{\pi d_F},$$
(3–28)

where η = the single-fiber efficiency,
   d_F = the diameter of the filter fiber,
   H = the depth of the filter,
   α = the volume fraction of solid fiber (1 − void fraction),
N_1 and N_2 = the concentrations of particles in the gas before and after the filter, respectively.

Until recently there were no means of estimating filter efficiency from basic considerations. Experimental measurements had to be made of N_1 and N_2 to compute the efficiency. In 1958 Friedlander [10] correlated data relating the single-fiber efficiency for glass-fiber filters of high porosity as follows:

$$\eta N_R N_{\text{Pe}} = 6(N_R N_{\text{Pe}}^{1/3} N_{\text{Re}}^{1/6}) + 3(N_R N_{\text{Pe}}^{1/3} N_{\text{Re}}^{1/6})^3,$$
(3–29)

where η = the efficiency,
   N_R = the direct interception parameter, d_p/d_{filter},
   N_{Pe} = the Peclet number, $d_{\text{filter}}u/D_{BM}$ (where u is the gas velocity and D_{BM} the particle diffusivity),
   N_{Re} = the Reynolds number, $d_{\text{filter}}u\rho_f/\mu_f$.

Use of the above equation is limited to values of N_{Re} and $N_I < 1$, where N_I is the inertial factor, $(mu/3\pi\mu d_p d_{filter}) < 1$. Although complex and limited in application, Friedlander's equation gives a basis for rational design in cases where experimental measurements are not available. The efficiency of a filter can be estimated by using Friedlander's equation to calculate the single-fiber efficiency, and then this value of η in the efficiency equation as follows:

Let $d_p = 10^{-5}$ ft (approximately 3 microns),
$$d_F = 3.5 \times 10^{-5} \text{ ft},$$
$$u = 3 \text{ ft/sec},$$
$$\mu_f = 1.2 \times 10^{-5} \text{ lb/(ft)(sec)},$$
$$\rho_f = 0.075 \text{ lb/ft}^3,$$
$$D_{BM} = 1.0 \times 10^{-10} \text{ ft}^2/\text{sec},$$
$$H = 0.01 \text{ ft},$$
$$\alpha = 0.05.$$

Then

$$N_R = \frac{10^{-5}}{3.5 \times 10^{-5}} = 0.286,$$

$$N_{Pe} = \frac{(3.5 \times 10^{-5})(3)}{10^{-10}} = 1.05 \times 10^6, \qquad (N_{Pe})^{1/3} = 101.7,$$

$$N_{Re} = \frac{(3.5 \times 10^{-5})(3)(7.5 \times 10^{-2})}{1.2 \times 10^{-5}} = 0.654, \qquad (N_{Re})^{1/6} = 0.954,$$

$$\eta = \frac{6(0.286 \times 101.7 \times 0.954) + 3(0.286 \times 101.7 \times 0.954)^3}{(0.286)(1.05 \times 10^6)} = 0.214.$$

Substitution of this value into the efficiency equation gives

$$\ln \frac{N_1}{N_2} = \frac{4(0.01)(0.05)(0.214)}{\pi(3.5 \times 10^{-5})} = 3.89,$$

or

$$\frac{N_1}{N_2} = 49.$$

Thus the filter efficiency $= (49 - 1)/49 = 98.0$ percent.

An equation for determining the pressure drop across a fiber filter of porosity less than 0.99 has been given by Chen [4] as

$$\Delta p = \frac{4}{\pi} \frac{K_4}{\ln K_5 \alpha^{-0.5}} \left(\frac{\alpha}{1-\alpha}\right) \left(\frac{\mu_f u_0 L_F}{d_{(f)s}^2}\right), \tag{3-30}$$

where K_4 = shape-factor constant,

$\quad K_5$ = orientation-factor constant,

$\quad \alpha$ = volume fraction of fibers in a filter,

$\quad \mu_f$ = gas viscosity,

$\quad u_0$ = superficial gas velocity,

$\quad L_F$ = thickness of filter,

$\quad d_{(f)s}$ = surface average fiber diameter.

Pressure drop across a fabric filter is a function of the resistance of the filter medium plus the resistance of the accumulated layer of particles. The flow through the fabric and layer is generally in the region of streamline flow, and the total pressure drop is then proportional to the gas velocity and the sum of the two resistances, given as

$$\Delta p = K_0 V_S + K_1 V_S w, \tag{3-31}$$

where K_0 = the fabric resistance coefficient in inches of water (ft/min),

$\quad K_1$ = the filter resistance coefficient for the particle in inches of water (lb dust/ft²) (ft/min of gas),

$\quad V_S$ = the filter ratio (ft³/min)/(ft²),

$\quad w$ = the weight of cake per unit fabric area in pounds per square foot.

Typical values of K_0 range from 0.01 to 0.04 [33], and K_1 values for common dusts range from about 0.5 to 2 for 20-mesh particles to 20 to 25 for 20-micron particles [52].

Wet Scrubbers

Wet scrubbers for removing particulate matter can be described as devices that use liquid to remove or aid in removing solids from the gas stream. Common types are spray towers, packed towers, cyclone scrubbers (Figure 3–10), jet scrubbers (Figure 3–11), venturi scrubbers, impingement scrubbers, and mechanical scrubbers (Figure 3–12). Some of the advantages of wet scrubbers are (a) moderately high efficiency in removing particles of 5-micron size and larger (some devices are capable of removing particles as small as 1 to 2 microns), (b) applicability to cleaning hot gases, and (c) moderate capital cost. The principal disadvantage is the high power consumption necessary to attain high collection efficiency.

Basically, the action occurring in a wet scrubber results from the intermingling of fine liquid and aerosol particles. Large solid particles are captured by interception and impingement, and small particles are entrapped by diffusion into the droplets. In addition, small particles may agglomerate when supercooled water condenses on their surface.

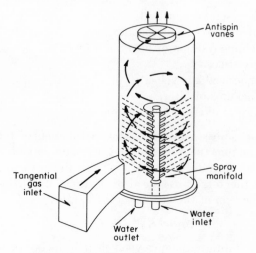

Figure 3–10 Wet cyclonic scrubber.

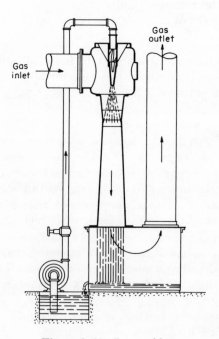

Figure 3–11 Jet scrubber.

Figure 3–12 Mechanical scrubber.

The efficiency of collection by impingement is a function of the dimensionless factor, $Dg/u_d u_f$, where u_f is the free-falling speed of the dust particle (which has a speed relative to the liquid droplet of u_d) and D is the droplet diameter [41]. A high collection efficiency requires a high relative speed between the gas and liquid—and a very small droplet diameter. These two conditions cannot be achieved in a gravity-spray tower, and the efficiency drops very sharply with decreasing particle size, as shown in Figure 3–13. This is achieved in certain scrubbers either by utilizing the moving gas to form the fine spray or by projecting the spray at high velocity toward the dust.

Spray towers consist simply of a chamber into which liquid enters at the top through spray nozzles, and the gas flows countercurrently from the bottom. These have been used frequently for cleaning the gas in municipal incinerators. The sprays are located either in a settling chamber

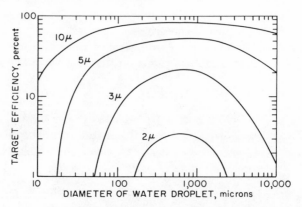

Figure 3-13   Efficiency of wet scrubbers.

or in additional chambers between the furnace and the stack breeching. A mist collector containing baffles usually follows the spray section to remove water droplets entrained by the gas. Gas velocities of 3 to 5 ft/sec are used in spray towers. High efficiency of removal is difficult to achieve for particles smaller than 5 microns.

In a cyclone scrubber the liquid is sprayed from a centrally located manifold at the top; the gas enters tangentially, usually at the bottom, and flows in a spiral pattern. The falling liquid droplets are hurled across the gas stream to the walls. Less entrainment of the liquid, higher throughput, and improved efficiency are obtained as compared with the simple spray tower. The superficial gas velocities in cyclone scrubbers range from 4 to 8 ft/sec. Water rates for spray towers and cyclone towers may vary from 1 to 20 gal per 1000 ft³ of gas, depending on the size of particles in the gas and the desired efficiency of removal. Cyclones are usually designed to remove particles of 2-micron and larger sizes. Their principal drawback is the relatively high pressure drop and correspondingly high power requirements.

A general equation for evaluating the efficiency of spray towers is [18]

$$\eta = 1 - \exp\left(\frac{-3\eta'qH}{2DG}\right) \tag{3-32}$$

where η' = the individual droplet efficiency,
   D = the diameter of the droplet in centimeters,
   q = the volumetric flow rate of liquid in cubic feet per second,
   G = the volumetric flow rate of gas in cubic feet per second,
   H = the distance through which the droplets travel in feet.

For cyclone scrubbers the efficiency equation is similar except that H is replaced by r, the radius of the cyclone.

Semrau [38] has developed a correlation of the efficiency applicable to several types of wet scrubbers based on the number of transfer units, N_T,

$$N_T = \alpha P_T \gamma,$$

where P_T is the contacting power, in horsepower per 1000 ft³ of gas per minute, expended in gas-liquid contacting in the scrubber; α and γ are constants characteristic primarily of the particle. The number of transfer units, N_T, is related to the collection efficiency η by

$$N_T = \ln\left(\frac{1}{1-\eta}\right).$$

Semrau's correlation is illustrated by the logarithmic plot of N_T with contacting power in Figure 3–14.

Jet scrubbers operate with a stream of water flowing at high velocity through a nozzle to induce a flow of gas and provide intimate mixing of the gas and liquid. They are highly effective for removing particles down to about 2 microns in size.

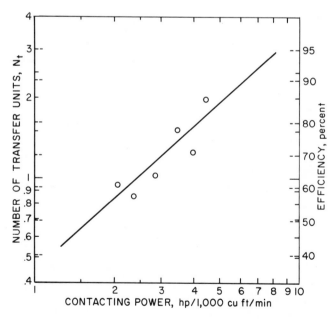

Figure 3–14 Contacting power and transfer units of wet scrubbers.

Impingement scrubbers attempt to provide increased efficiency by maintaining a high relative velocity between the spray droplets and the smallest dust particles in the gas. These scrubbers consist of a series of bubble plates with target plates above the openings in the bubble plates to restrict the flow. Gas flows through holes in the bubble plates and strikes the target plate, which is wetted or submerged by the liquid. Fine spray is produced by the gas passing through the restricted openings around the edges of the targets. Stairmand [41] indicates that removal efficiencies of up to 80 percent for 1-micron particles can be obtained in properly designed impingement scrubbers. The use of impingement scrubbers to reduce emissions from flue-fed incinerators is described in a publication by Kaiser et al. [19].

There are various types of mechanical scrubbers in which the liquid is sprayed onto rotating blades or disks. One commonly used type consists of alternate rows of rotor and stator arms. The water, which is injected axially, is converted to a fine spray of droplets with high velocity. Good efficiency is reported for removing particles down to 1 micron in size [17], but power consumption is high.

Electrostatic Precipitators

Electrostatic precipitators are the most satisfactory device for removing small particles from moving gas streams at high collection efficiencies. They have been used almost universally in power plants for removing fly ash from the gases prior to discharge. However, electrostatic precipitators have been used very little in treating gases from incinerators because of their high cost and the varying nature of the electrical resistivity of the particulate matter [23]. Separation of suspended particles is accomplished by the attraction of electrically charged particles to a charged surface. The charge on the particles is related to the difference in the dielectric constant between the particles and gas.

A precipitator basically consists of a discharge electrode (usually negatively charged) that provides the initial source of electrons for charging the particles and collecting electrodes that provide a surface for collecting and holding the dust. A corona discharge is maintained in the space between the electrodes by imposing a voltage of from 25 to 100 kV. This applied voltage must be sufficient to cause free electrons in the gas to strike the gas molecules with enough force to release additional electrons. Because the molecules become ionized, the gas is conducting and given a sustained discharge characterized by a visible glow. Other components that are necessary in an electrostatic precipitator are a high-voltage power supply and a means of removing the collected dust, such as mechanical rapping devices.

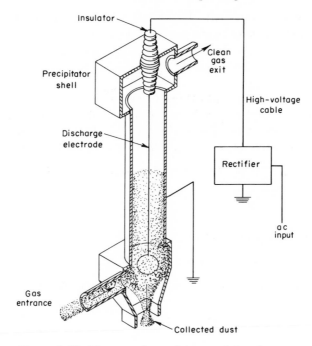

Figure 3–15 Electrostatic precipitator of the pipe type.

Precipitators are of two types: pipe and plate. In the former vertical parallel pipes serve as collecting electrodes and wire or rods located axially serve as discharge electrodes (Figure 3–15). The gas flows through the pipes, and the particles collect on the inner wall. The collecting electrodes in the plate-type precipitator are parallel plates either solid metal or lattice structured. The discharge electrodes are similar to those used in the pipe type and are located midway in the passage between the plates through which the gas flows. Lagarias [20] has described the various types of discharge electrodes in common use and the factors that affect the selection of each type.

The collection efficiency of an electrostatic precipitator can be calculated from an equation derived by Deutsch [7] and modified by White [48] as follows:

$$\eta = 1 - e^{-(A_c W/q)}, \tag{3-33}$$

where A_c = the collecting surface area,

W = the drift-velocity constant (velocity component of the particle in the direction of the collecting electrode),

q = the volumetric gas flow.

The drift-velocity constant, W, is related to the particle size, field strength, and properties of the gas, as defined by

$$W = \left[1 + \frac{2(K_e - 1)}{K_e + 2} \right] \frac{rE_0^2}{6\pi\mu_f}, \tag{3-34}$$

where K_e is the dielectric constant of the particle.

For a plate precipitator of length L, height H, and spacing s and for a given volumetric flow of gas q, linear velocity u and time t for the gas within the active plate surface, the following applies [49]:

$$A_c = 2LH \text{ (two surfaces per space)},$$

$$q = Hsu = \frac{HsL}{t}. \tag{3-35}$$

The efficiency equation of Deutsch [7] becomes

$$\eta = 1 - e^{-2tW/s}. \tag{3-36}$$

Similarly, for a pipe-type precipitator of radius r, length L, and gas velocity u

$$A_c = 2\pi rL,$$

$$V = \pi r^2 u = \pi r^2 \frac{L}{t},$$

and

$$\frac{A_c}{q} = \frac{2t}{r}.$$

Thus the collection efficiency can be written for pipe-type precipitators as

$$\eta = 1 - e^{-2tW/r}. \tag{3-37}$$

The equations for the two types of precipitators are nearly alike except that for the plate-type precipitator the distance s is the perpendicular distance between collecting surfaces, whereas for the pipe type the radius r is the distance from the discharge electrode to the collecting surface.

Once the limiting size of particles to be removed is fixed, the three items in the design of the precipitator that can be varied are the field strength E_0; the plate spacing s or pipe radius r; and the time t, which varies inversely with the gas velocity. The practical range of gas velocities is from 3 to 20 ft/sec.

Although it is based on the theoretical considerations, the design of electrostatic precipitators still relies on much empirical "know how." The designer must obtain the proper drift velocity from actual field experience in many installations or experimentally. Lagarias [21] plotted the effi-

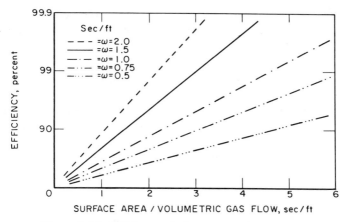

Figure 3–16 Efficiency of electrostatic precipitators.

ciency with change in A_c/q (Figure 3–16) with drift velocity as parameter. At the chosen drift velocity the ratio of collection-surface area to volumetric gas flow required for a desired efficiency of removal can be read directly. White and Cole [50] describe the factors to be considered in designing a precipitator for a new application to obtain a 99.8-percent collection efficiency of a 0.7-micron oil fume at a high gas velocity of 25 ft/sec.

As mentioned earlier in this section, there has been little application of electrostatic precipitators to incinerator gases because of the high cost and nonuniformity of the emissions. This latter item is significant because of the range of electrical resistivities of the various particulates. Materials of low resistivity, such as carbon black, are readily precipitated but lose their charge at the collecting electrode, do not adhere, and are readily re-entrained in the gas. On the other hand, materials of high resistivity are precipitated with more difficulty and can act as an insulating coat on the collector, thus impairing the operation of the precipitator. Sproul [40] indicates that the resistivity should be less than 10^{11} ohm-cm for good performance, and White [48] states that the optimum is 2×10^{10} ohm-cm.

Sonic and Ultrasonic Separation

Ultrasonic devices to separate particulate matter or to augment other separating techniques have been used for many years. Several ultrasonic separators were constructed shortly after World War II, but the use of many has been discontinued because the power consumption was excessive. Boucher [1] has presented an excellent review of the status of ultrasonics and the new developments that have effected significant improvement in the field. A brief discussion of ultrasonics is included because of its potential applications for separating the fine dispersoids present in incinerator emissions.

An airborne sound wave consists of a compression wave followed by a rarefaction wave traveling in a gaseous medium. A sound wave of suitable frequency and sufficient intensity that is transmitted through a gas containing particulate matter creates turbulence, which causes collisions and agglomeration of the solids. This principle is applied practically by the use of a tower or agglomeration chamber through which the dust-laden gas flows and in which an intense ultrasonic standing-wave field is maintained. The tower serves as a primary collector, and closely adjacent to the tower is a cyclone, filter, or scrubber to remove the agglomerated solids. A typical installation is shown in Figure 3–17.

The two important items that influence the effectiveness of the agglomeration are the characteristics of the sound generator and the overall design of the unit. Static generators (whistles) or dynamic generators (sirens) produce ultrasonic waves in gases. The static type employs a gas jet directed at high speed at a barrier or into a resonance chamber, whereas the dynamic generator consists of a mechanically driven perforated rotor rotating at 8000 to 20,000 rpm atop a stator containing many small circular openings. On rotation, passages are opened and closed extremely rapidly, and the flow of compressed gas through these openings produces an intense acoustic wave. Usually the frequency of sirens of this type ranges from 1.5 to 25 kHz.

Boucher [1] points out that the factors affecting acoustic agglomeration are the sound intensity, frequency, exposure time in the sonic field, aerosol concentration, gas temperature, and aerosol turbulence. These factors are utilized in the design of separating equipment, taking into account aerodynamic, structural, and economic considerations as well. Pilot-plant data must be available before a large unit can be designed. Inoue [15] has given the following general approach for designing ultrasonic devices:

1. Calculate the required intensity I from the equation

$$I = \frac{4\beta I_0}{\pi D^2},\tag{3-38}$$

where I = the average intensity in watts per square centimeter,
   I_0 = the acoustic power delivered by the generator in watts,
   β = a constant that is characteristic of the unit,
   D = the tower diameter in centimeters.

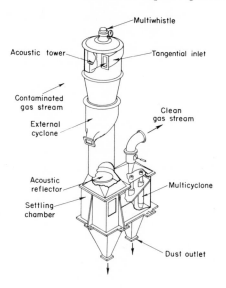

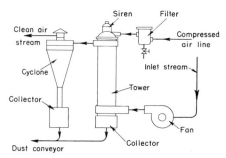

Figure 3–17 Typical tower installation with acoustic device.

2. Calculate the coalescence constant C,

$$C = \sqrt{I} \tag{3–39}$$

and the agglutination index I_A:

$$I_A = \frac{d_p}{d_0} = e^{Ct/3}, \tag{3–40}$$

where d_0 and d_p are the mean particle diameters before and after the sonic exposure, respectively, and t is the contact time in seconds of the gas in the tower.

3. Calculate the required height of tower L (in centimeters):

$$L = \frac{3u \ln (d_p/d_0)}{C}, \tag{3-41}$$

where u is the mean gas velocity in centimeters per second and C is the coalescence constant obtained in calculation 2. A usual value of the gas velocity is 100 cm/sec.

Most of the recent commercial applications of ultrasonic cleaning have been in Europe. Fumes from metallurgical operations, tar mists, phthalic anhydride effluent, and even power-plant exhaust gases are being treated with good results. Investigations are under way on the use of ultrasonics to supplement other devices, such as sonic emissions to clean dust particles from bag filters, and to increase the efficiency of cyclones.

The attractive features of ultrasonic devices are the low capital cost and applicability to gases at elevated temperatures. However, power consumption is generally excessive, a value of 4 kW per 1000 cfm of gas treated being common. Operating costs this high are acceptable only in special cases.

ODOR CONTROL

The emission of malodorous components in incinerator exhaust can cause a serious problem. There are periods in both municipal and flue-fed incinerators during which the furnace temperature is too low to effect complete combustion of the organic compounds that are volatilized from the refuse. Kaiser et al. [19] measured the emissions of particulates and noxious gases from a flue-fed incinerator operating at different burning and air rates; the content of noxious gases ranged from about 0.9 to 3 lb per 100 lb of refuse. The average odor concentrations ranged from 2.5 to 100 ASTM odor units. An ASTM odor unit of 1.0 represents a gas with a barely perceptible odor; thus a gas of 2.0 odor units requires dilution with an equal volume of odor-free air to be barely perceptible. Surprisingly no correlation was found between odor level and concentration of noxious gases. However, there was a fairly good relationship of the odor intensity to the aldehyde emission. Discussed briefly in this section are the basic fundamentals of the common methods of odor control and their possible application to incinerator gases. Turk [45], who has covered the field of odor control thoroughly, points out that odor threshold concentrations are usually several-fold lower than those of toxicity, irritation, and plant damage. Consequently odor control requires efficient methods of removal, often to negligible residual concentration, to achieve the desired results. As an alternative to these measures it is sometimes possible to avoid the formation or emission of odorous materials by process modifications. In the case of incinerators these could be modifications in the design of equipment or in the firing practices.

Dilution and Dispersal

Dilution of contaminated gas with clean air or other gases does not decrease the total quantity of odorous compounds discharged, but it does reduce the odor intensity. Since the usual method of dispersal is via a stack to the atmosphere, meteorological conditions must be considered in discharging odorous gases in this manner. The odor intensity is proportional to the logarithm of the odorant concentration; however, since the odor threshold concentration may be quite low, considerable dilution may be required to decrease the odor to a satisfactory level to avoid possible localized pollution. Improved dispersion is achieved if the diluting air pumped into the exhaust stack increases the upward velocity at the exit, thereby increasing the effective stack height [53]. A similar effect is achieved by increasing the exit temperature of the gas. Odor control by dispersal is questionable not only because it is closely related to meteorological conditions but also because odors may persist for some distance after the noxious material is discharged from the stack.

Adsorption

Adsorption is the process whereby molecules of gases or liquids are retained on the surface of solids. As applied to odor control, the gas or air stream containing the odorous compounds is passed through a bed of sorbent, usually activated carbon, and the odorous gases and vapors are retained. After a given exposure the adsorbent approaches saturation and loses its effectiveness, and a breakthrough of the odorant tends to occur. Before this happens the spent bed is taken off stream, and a fresh or reactivated bed is substituted. The spent adsorbent may be discarded but is usually reactivated by heating or by a solvent. Activated carbon is reactivated with superheated steam at 1000 to 1300°F.

The design of adsorption equipment involves consideration of gas throughput, capacity of the adsorbent, allowable pressure drop, allowable linear velocity of the gas, and service life between reactivations. The service life between reactivations can be computed from the following relationship [45]:

$$t = \frac{6.43 \times 10^6 S_r w}{E_f Q_r M C_v}, \tag{3-42}$$

where t = service life in hours,

   S_r = retentivity or ultimate proportionate saturation of carbon,
   w = weight of adsorbent,
   E_f = fraction adsorption efficiency,
   Q_r = air flow in cubic feet per minute,
   M = average molecular weight of contaminants,
   C_v = concentration of contaminants in parts per million.

Turk states that a common design of commercial equipment using a thin bed of activated carbon at 95 percent removal efficiency operates at 700 ft^3/(min)(cell) containing 45 lb carbon. In this design $t = 4.35 \times 10^5 \, S_r/MC_v$. For removal of 2 ppm of acrolein (molecular weight = 56 and $S_r = 0.15$) the service time is then about 580 hr. Since the service life is inversely proportional to the concentration of the odorous compounds, adsorption is generally limited to removal of low concentrations, in the range of 2 to 5 ppm. In addition to activated carbon, molecular sieves have been suggested for removing odorants, but their use has been limited because they adsorb water preferentially, which reduces their effectiveness for adsorbing odorants.

Combustion

Both noncatalytic and catalytic combustion are employed to eliminate odorous compounds, the latter usually when the concentration of combustibles is low. The combustion device must provide complete combustion, because, as in the case of the incinerator, the products of incomplete combustion are malodorous. High temperature, turbulence, adequate residence time, and sufficient oxygen are the elements that are required for complete combustion [28]. In noncatalytic firing the required temperature for complete combustion is about 1100 to 1500°F. To achieve such temperatures a high consumption of auxiliary fuel may be necessary.

Ludwig [26] and Orning et al. [29] independently studied oxidation of low concentrations of combustion intermediates from incineration. Using acetaldehyde and carbon monoxide as his reactants, Ludwig found that even at 1200°F about 10 to 20 percent of these compounds remained unconverted with 40 percent excess air and a residence time of 0.3 to 0.5 sec. Use of high excess-air quantities of 100 and 200 percent gave less conversion than use of 40 percent excess. High turbulence was a key factor in obtaining high conversion. Orning and co-workers studied several types of organic compounds, including aliphatic and ring hydrocarbons, organic acids and bases, esters, ketones, and alcohols. Except for benzene and acetic acid, the compounds were destroyed at 1292°F in the presence of 1 to 4 percent oxygen within 0.3-sec residence time.

In a few cases complete reaction to carbon dioxide and water did not occur; instead some quantity of compounds more stable than the original material was formed that still might cause an odor problem.

Use of a catalyst accelerates oxidation so that the required combustion temperature need be only 500 to 800°F [13]. Catalytic oxidation occurs without flame on the catalyst surface. The heat liberated by the exothermic reaction causes a rise in the gas temperature. If the gas to be treated is at low temperature, it may be heat exchanged with the exit gas to raise the incoming gas to the required temperature for oxidation. In several installations sufficient excess heat has been recovered as waste-heat steam to more than offset the cost of the installation.

Most common catalysts are of the platinum group or a mixture of non-noble metal oxides impregnated on grid supports to keep the pressure drop low [32]. The major operating cost is that of the catalyst replacement due to loss of activity. Principal reasons for loss of activity are contamination by catalyst poisons, physical blockage of the surface by deposit of particulate matter or by carbon formed on incomplete combustion, and mechanical loss of the catalyst by abrasion and erosion. The cost of catalytic combustion must be computed for each case individually and depends principally on the combustible content of the gas, the catalyst life realized, the amount of heat recovery and value of recovered heat, and the fixed capital costs.

Odor Modification or Masking

Odor modification is the means of counteracting odors by the addition of nonreactive vapors. Ideally the odor intensity may be decreased to zero, a condition known as cancellation. More frequently counteraction effects a sufficient decrease in the odor intensity of the objectionable components to an acceptable level or results in a more pleasant odor than the original. Odor modification is often difficult to evaluate because both the quality and the intensity of the odor may be altered. The effect of counteractants is usually evaluated by tests with a panel. Some of the materials used as counteractants are surfactive agents, nonsurface-active sulfonates, chelate-type oxygen absorbent compounds, sulfamates, aldehydes, ketones, olefins, conjugated diolefins, and acetylene [30]. Despite its empirical nature, odor modification is successfully used in many difficult applications, such as sulfate pulp mills and rendering plants.

Wet Scrubbing

Scrubbing with liquids has been employed for absorption of odorous materials from incinerators, often with simultaneous removal of part of the particulate matter. Absorption of odorants may occur with or without chemical reaction with the liquid agent. Physical solubility without chemical reaction takes place as long as the partial pressure of the solute in the liquid phase is below the equilibrium value. Equilibrium determines the limit of absorption. The other important factor determining the quantity dissolved is the rate of absorption. Absorption-rate constants are determined experimentally. Water is the preferred solvent, but many of the odorants are insoluble in water and require fairly costly solvents.

To obtain efficient removal by scrubbing good contact between the gas and liquid must be obtained. Most of the scrubbers already mentioned for removing particulate matter are also applicable to washing odorous gases. The principal types of scrubbers are spray tower, packed tower, plate tower, venturi, and jet type. The use of water scrubbing in spray or cyclone towers to treat gases from flue-fed incinerators has been already mentioned. Byrd and Dewey [3] describe the application of a venturi scrubber for overcoming an odor problem in exhaust gases.

NOMENCLATURE

Symbol		
A	cross-sectional area	
$\acute{A}$	percentage of adhesion	
A_c	collecting-surface area	
B_c	diameter of cyclone inlet	
C	coalescence constant	
C_D	drag coefficient	
C_v	concentration of contaminants	
d_0	mean particle diameter of solids before exposure	
d_F	diameter of filter fiber	
d_{filter}	filter diameter	
$d_{(f)s}$	surface average fiber diameter	
d_p	particle diameter	
d_p'	particle diameter, larger particle	
D	diameter of tower, cylinder, or droplet	
D_{BM}	particle diffusivity	
D_c	cyclone diameter	
D_e	diameter of cyclone gas outlet	
e	charge per electron	
E	field strength	
E_0	electric-field strength	
E_f	fraction adsorption efficiency	
E_P	collecting-field strength	
f	force (drag force, force of attraction, etc.)	
F_{cv}	number of inlet velocity heads	
g	gravitational constant	

G	volumetric flow rate of gas in cubic feet per second	
h	height of travel	
H	height or depth	
H_c	height of cyclone inlet duct	
I	average intensity	
I_0	acoustic power delivered by generator	
I_A	agglutination index	
k	Stokes'-law coefficient	
K	Cunningham constant	
K_0	fabric resistance coefficient	
K_1	filter resistance coefficient	
K_4, K_5	shape and orientation factors of filters	
K_B	Brownian-agglomeration constant	
K_c	cyclone constant	
K_e	dielectric constant	
K_S	velocity-agglomeration constant	
L	length	
L_F	thickness of filter	
m	mass of particle	
M	molecular weight	
n	number of particles per cubic centimeter	
n'	number of electrons	
n_0	initial number of particles	
N	Avogadro's number	

N_1, N_2	concentrations of particles before and after filter	u_{ct}	centrifugal-settling velocity
N_I	inertial factor	u_d	droplet velocity
N_{Pe}	Peclet number	u_f	free-falling speed of dust particle
N_R	interception parameter	u_g	vertical velocity due to gravity
N_{Re}	Reynolds number	u_r	radial velocity of particle relative to the gas
N_s	number of spirals made by gas in cyclone	u_t	terminal velocity
N_t	number of transfer units	u_{tan}	tangential velocity
p	factor in charge on particle $= 3E/(\epsilon + 2)$	V	volume
		V_p	particle volume
P	particle parameter for impingement efficiency	V_S	filter ratio
		w	weight of cake per unit fabric area or weight of adsorbent
P_T	contacting power	W	drift-velocity constant
q	volumetric rate of flow	x	average displacement of particles
q_0	number atoms per cubic centimeter	X	distance between particles
Q	electric charge	α	volume fraction of solid fiber (1 − void fraction)
Q_0	limiting electric charge	α, γ	constants
Q_r	air flow	β	constant for acoustic generator
r	radial distance, radius of cyclone or particle	ϵ	dielectric constant
R	gas constant	η	efficiency of collision, filter, droplet, or removal
s	spacing	η'	individual-droplet efficiency
S	average surface roughness	λ	mean free path of fluid molecules
S_r	retentivity	λ'	London-van der Waals constant
t	time, service life	μ_f	fluid viscosity (liquid or gas)
T	absolute temperature	ρ_f	fluid density (liquid or gas)
u	velocity	ρ_p	particle density
u_0	superficial gas velocity	$\Delta\rho$	$\rho_p - \rho_f$
u_c	average gas velocity at cyclone inlet	Δh	inlet velocity head
		Δ_p	pressure differential

REFERENCES

[1] Boucher, R. M. G., "Ultrasonics in Processing," *Chem. Eng.*, **68**, No. 20, 83–100 (October 2, 1961).

[2] Bradley, R. S., "The Cohesive Forces Between Solid Surfaces and the Surface Energy of Solids," *Phil. Mag.*, **13**, 853–862 (1932).

[3] Byrd, J. F., and E. L. Dewey, "Venturi Scrubber in Odor Control," *Chem. Eng. Progr.*, **53**, No. 9, 447–451 (September 1957).

[4] Chen, C. Y., "Filtration of Aerosols by Fibrous Media," *Chem. Reviews*, **55**, 595–623 (1955).

[5] Corn, M., "The Adhesion of Solid Particles to Solid Surfaces, II," *J. APCA*, **11**, No. 12, 566–584 (December 1961).

[6] Cunningham, E., "On the Velocity of Steady Fall of Spherical Particles through Fluid Medium," *Proc. Roy. Soc.* (London), **83**, 357 (1910).

[7] Deutsch, W., "Motion and Charge of a Charged Particle in the Cylindrical Condenser," *Ann. Physik*, **68**, 335–344 (1922).

[8] Einstein, A., "Elementary Theory of Brownian Movement," *Z. Elektrochem*, **13**, 41–42 (1907).

[9] Flood, L. P., "Air Pollution Resulting from Incineration—Its Reduction and Control," *J. APCA*, **9**, No. 1, 63–68 (May 1959).

[10] Friedlander, S. K., "Theory of Aerosol Filtration," *Ind. Eng. Chem.*, **50**, 1161–1164 (1958).

[11] Hamaker, H. C., "The London-van der Waals Attraction Between Spherical Particles," *Physica*, **4**, 1058–1072 (1937).

[12] Herne, H., "The Classical Computations of the Aerodynamic Capture of Particles by Spheres," *Int. J. of Air Pollution*, **3**, Nos. 1–3, 26–34 (October 1960).

[13] Houdry, E. J., "Practical Catalysis and Its Import on Our Generation," in *Advances in Catalysis*, Hugh Taylor, ed., Academic, New York, 1957, pp. 481–487.

[14] Hovey, H. H., A. Risman, and J. F. Cunnan, "The Development of Air Contaminant Emission Tables for Nonprocess Emissions," *J. APCA*, **16**, No. 7, 362–366 (July 1966).

[15] Inoue, I., "The Sonic Agglomeration Apparatus. Sound Generator and Agglomeration Tower," *J. Chem. Ind. Japan*, **18**, 180–186 (1954).

[16] Jackson, R., "Settling Chambers for the Collection of Grit," *Review No. 189, BCURA*, **23**, No. 9, Part II, 349–365 (August-September 1959).

[17] Jennings, R. F., "Blast Furnace Cleaning," *J. Iron Steel Inst.*, **164**, 305–325 (1950).

[18] Johnstone, H. F., and M. H. Roberts, "Deposition of Aerosol Particles from Moving Gas Streams," *Ind. Eng. Chem.*, **41**, No. 11, 2417–2423 (November 1949).

[19] Kaiser, E. R., J. Halitsky, M. D. Jacobs, and L. C. McCabe, "Modifications To Reduce Emission from Flue Fed Incinerator," *J. APCA*, **10**, 183–197; 207; 251 (1960).

[20] Lagarias, J. S., "Discharge Electrodes and Electrostatic Precipitators," *J. APCA*, **10**, No. 4, 271–274 (August 1960).

[21] Lagarias, J. S., "Predicting Performance of Electrostatic Precipitators," *J. APCA*, **13**, No. 12, 595–599 (December 1963).

[22] Lapple, C. E., "Fine Particle Characteristics" *Stanford Research Institute Journal*, **5**, 95, Third Quarter, 1961.

[23] Lenehan, J. W., "Air Pollution Control in Municipal Incineration," *J. APCA*, **12**, No. 9, 414–417 (September 1962).

[24] Leniger, H. A., Chapter 1 in "Phase Separation and Classification," *Cyclones in Industry*, K. Rietema and C. G. Verver, eds., Elsevier, Amsterdam, 1961, p. 151.

[25] Loeb, L. B., *Kinetic Theory of Gases*, McGraw-Hill, New York, 1934.

[26] Ludwig, J. H., "Thermal Destruction of Low Concentrations of Acetaldehyde Vapor and Carbon Monoxide," Paper 58–15, 51st Annual Meeting APCA, Philadelphia, May 25–29, 1958.

[27] McCormick, P. Y., R. L. Lucas, and D. F. Wells, "Gas-Solid Systems," Section 20, *Chemical Engineer's Handbook*, 4th edition, R. H. Perry, C. H. Chilton and S. D. Kirkpatrick, eds., McGraw-Hill, New York, 1963, pp. 68–74.

[28] Mills, J. L., W. F. Hammond, and R. C. Adrian, "Design of Afterburners for Varnish Cookers," *J. APCA*, **10**, No. 2, 161–168 (April 1960).

[29] Orning, A. A., J. J. Pfeiffer, W. C. Harrold, and J. F. Shultz, "A Progress Report on an Experimental Study of Incineration," Paper No. 58–12; 51st Annual Meeting APCA, Philadelphia, May 25–29, 1958.

[30] Pantaleoni, R., "An Approach to Odor Control," *J. APCA*, **5**, No. 4, 213–215 (February 1956).

[31] Pilpel, N., "Industrial Gas Cleaning," *Br. Chem. Eng.*, **5**, 542–550 (August 1960).

[32] Prass, H. J., "Waste Gases Can Be Cleaned at Low Cost," *Power*, **107**, 74–77 (April 1963).

[33] Pring, R. T., "Data from Bag-Type Cloth Dust and Fume Collectors," *Air Pollution*, L. C. McCabe, ed., McGraw-Hill, New York, 1952.

[34] Rose, A. H., Jr., R. L. Stenburg, M. Corn, R. R. Horsley, D. R. Allen, and P. W. Kolp, "Air Pollution Effects of Incinerator Firing Practises and Combustion Air Distribution," *J. APCA*, **8**, 297–309 (February 1959).

[35] Rose, A. H., Jr., D. G. Stephan, and R. L. Stenburg, "Prevention and Control of Air Pollution by Process Changes or Equipment," *Air Pollution*, World Health Organization Monograph Series No. 46, Geneva, 1961, pp. 407–434.

[36] Rosin, P., E. Rammler, and W. Intelmann, "Principle and Limits of Cyclone-dust Removal," *Z. Ver. dtsch. Ing.*, **76**, 433–437 (1932).

[37] Sell, W., "Dust Separation on Air Filters," *Forsch. Gebiete Ingenieurw.*, **2**, Forschungsh. No. 347, Aug. 1931

[38] Semrau, K. T., "Correlation of Dust Scrubber Efficiency," *J. APCA*, **10**, No. 3, 200–207 (June 1960).

[39] Shepherd, C. B., and C. E. Lapple, "Flow Pattern and Pressure Drop in Cyclone Dust Collectors," *Ind. Eng. Chem.*, **31**, 974–984 (1939).

[40] Sproul, W. T., "Collecting High Resistivity Dusts and Fumes," *Ind. Eng. Chem.*, **47**, No. 5, 940–944 (May 1955).

[41] Stairmand, C. J., "The Design and Performance of Modern Gas-Cleaning Equipment," *J. Inst. of Fuel*, **29**, No. 181, 58–73 (February 1956).

[42] Stenburg, R. L., R. P. Hangebrauck, D. J. von Lehmden, and A. H. Rose, Jr., "Field Evaluation of Combustion Air Effects on Atmosphere Emissions from Municipal Incinerators," *J. APCA*, **8**, 297 (February 1959).

[43] Ter Linden, A. J., "Investigations into Cyclone Dust Collectors," *Proc. Inst. Mech. Eng.*, **160**, 233–251 (December 1949).

[44] Teverovskii, E. N., "Coagulation of Aerosol Particles in a Turbulent Atmosphere," *Akad. Nauk. Ser. Geogr. Geofiz.*, **12**, No. 1, 87–94 (1948).

[45] Turk, A., "If Your Problem Is Getting Rid of Obnoxious Odors, There Are Five Basic Methods Available," *Industrial Wastes*, **3**, pp. 9–14 (January–February 1958).

[46] Underwood, G., "Removal of Sub-Micron Particles from Industrial Gases, Particularly in the Steel and Electrical Industries," *Int. J. Air and Water Pollution*, **6**, 229–263 (1962).

[47] Von Smoluchowski, M., "Displacement of Particles by Molecular Bombardment," *Bull. Int. Acad. Sci. Cracovie*, 1906, p. 577.

[48] White, H. J., *Industrial Electrostatic Precipitation*, Addison-Wesley, Reading, Mass., 1963.

[49] White, H. J., "Modern Electrical Precipitation," *Ind. Eng. Chem.*, **47**, No. 5, 932–940 (May 1955).

[50] White, H. J., and W. H. Cole, "Design and Performance Characteristics of High-Velocity, High-Efficiency Air Cleaning Precipitators," *J. APCA*, **10**, No. 3, 239–245 (June 1960).

[51] Whytlaw-Gray, R., and H. S. Patterson, *Smoke*, Edward Arnold Co., London, 1932.

[52] Williams, C. E., "Determination of Cloth Area for Industrial Filters," *Heating, Piping, Air Conditioning*, **12**, 259–263 (1940).

[53] Witheridge, W. N., "Odor Control Practices," in *Air Pollution*, L. C. McCabe, ed., McGraw-Hill, New York, 1952.

[54] Zenz, F. A., and D. F. Othmer, *Fluidization and Fluid Particle Systems*, Reinhold, New York, 1960.

4

ON-SITE INCINERATION
OF RESIDENTIAL WASTES
A. DOMESTIC DWELLINGS

Glen M. Hein and Richard B. Engdahl

The simplicity and convenience of on-site incineration of domestic wastes has always made it an attractive method of waste disposal. The problems of on-site storage, collection, and transport of odorous wastes are few. Years ago the wood-fired stove often served for incineration of food wastes. Backyard burning is still widely practiced, especially in rural areas, but the smoke, odor, and unsightly appearance of most outdoor burners have forced their abandonment in the crowded suburbs and cities. Many outdoor burners have been replaced with household incinerators connected to the chimney. The first of these was patented in 1915 [7]. During the next 40 years residential incinerators came into wider use but did not evolve very far.

Project Leader, Thermal Systems, (deceased) and Fellow, Mechanical Engineering Department, respectively, Battelle Memorial Institute, Columbus, Ohio.

SIMPLE HOUSEHOLD INCINERATOR

For many years a typical incinerator consisted of a vertical, cylindrical, or box-shaped refuse-burning chamber mounted above an ash drawer, with a grate forming all or part of the bottom of the chamber. The charging door, usually located in the top, was either single or double walled. Most incinerators had an annular space between the sides of the combustion chamber walls and the insulated jacket walls. This space allowed air to circulate through the annulus from the bottom up and out through the top, thus cooling the jacket. Some incinerators discharged this cooling air into the room, whereas others fed it into the top of the incinerator for refuse combustion or flue-gas cooling. In one design this air flow was divided with some air directed into the combustion chamber and the rest drawn into the flue outlet just above the combustion chamber for flue-gas cooling [7].

Some incinerators had passageways of perforated cast iron inside the flue extending vertically upward from the grate level to the flue collar, whereas others had heavy-gage, expanded-metal screens inside the flue. One design used a solid-metal baffle to extend the flue passageway to below the grate level. This arrangement routed the flue products down through the grates before they entered the flue passageway to travel upward to the outlet. Thus each design provided a path, which presumably could not be obstructed by refuse, for flow of flue products from the interior of the incinerator either directly upward to the flue collar or indirectly downward through the grate and then up to the flue-outlet collar.

Many of these simple units used gas burners to ignite, dry, and burn the charge. The high-input type had the burner below, above, or at the grate level; whereas the dehydrator type (or low-input) incinerator usually had the burner in a housing above the grate. Multiple drilled ports or the single-port burner with various gas-input rates were predominant. However, a burner might have a target for deflecting the flame or a shell (or shroud) for protecting the flames and burner ports against fly-ash contamination or physical smothering by the charge.

In one survey on these simple units [8] the three most common complaints were pilot-light outage, burner outage, and improper operation. The causes of these troubles may be summarized as (a) ashes around the burner, (b) overloading the charging compartment, (c) not burning the charge fast enough, (d) not emptying the ashes, (e) clogging the grates with foil or other unburnable materials, and (f) improper charging and maintenance.

In nearly all instances owners of incinerators were pleased with them.

A few complaints were received from neighbors—but with all of these offending units the troubles were corrected by proper charging of garbage and papers, and regular shaking of grates and removal of ashes.

DEVELOPMENT OF SMOKELESS-ODORLESS INCINERATORS

As a result of surveys on the shortcomings of the conventional incinerators the American Gas Association (AGA) commissioned two laboratories to research and develop prototypes of smokeless, odorless, fly-ash-free units with improved control of flue-gas temperatures during the combustion of fast-burning, dry materials. As part of the program considerable study was given to establishing a test charge that would fairly represent domestic wastes [7].

Domestic Refuse—Composition and Quantity

In arriving at a representative daily load figure information was assembled from various sources on the amount of food refuse and combustible material per person per day in several municipalities. The data varied over a broad span as they generally included commercial and possibly some industrial wastes. The median figure for the United States represented at that time a span of 587 to 1575 lb/(capita)(yr) for garbage and 173 to 537 lb/(capita)(yr) for rubbish.

Table 4A-1 represents municipal-refuse analyses from two sources. Studies show that there are variations, depending on the locality being sampled, the time of year, and certain qualifications made at the time of sampling and analyzing. In using these analyses to establish domestic-incinerator charges one must remember that they have been made on a municipal basis and therefore are not exactly representative of household

Table 4A-1   Analyses of Municipal Refuse

Refuse	Moisture Content (percent)	Volatile Matter (percent)	Fixed Carbon (percent)	Ash Content (percent)	Heating Value (Btu/lb)	Average Weight (lb/ft³)	Production (lb/ capita) (day)
Garbage[a]	73.26	16.89	4.71	5.14	2233	40	—
Rubbish[a]	5.78	65.66	14.69	13.87	6832	7	—
Garbage[b]	65.00	22.00		13.00	1936	46	0.68
Rubbish[b]	7.00	79.00		14.00	7460	15	0.14

[a] Data from [1].
[b] Data from [2].

wastes because of the inclusion of large amounts of commercial refuse. However, the two analyses can serve in selecting a practical test charge. Because of the many variables involved, it is impossible to obtain a single analysis that can be considered representative for all parts of the country at all seasons or to duplicate all possible field conditions in a single test charge and procedure.

A more recent study of municipal refuse concluded that its composition is approximately as follows [5]:

Category	Percent
Rubbish	64
Food waste	12
Ash, glass metal, etc.	24

The great range of heat value of municipal refuse shown in Table 4A–1 is of course a direct result of the natural variation in heat value among the different materials commonly found in refuse (see Table 4A–2).

After careful consideration of these data it appeared that for residential purposes a representative average food waste per person per day

Table 4A–2   Heating Value of Various Substances[a]

Substance	Heating Value (Btu/lb, dry)	Substance	Heating Value (Btu/lb, dry)
Petroleum coke	15,800	Oats	7,998
Wood sawdust:		Wheat	7,532
Pine	9,676	Oil:	
Fir	8,249	Cottonseed	17,100
Rags:		Lard	16,740
Silk	8,391	Olive	16,803
Wool	8,876	Paraffin	17,640
Linen	7,132	Fats (animal)	17,100
Cotton	7,165	Butter	16,560
Cotton batting	7,114	Casein	10,548
Corrugated-fiber		Egg white	10,260
carton	5,970	Egg yoke	14,580
Newspaper	7,883	Candy	8,046
Wrapping paper	7,406	Pecan shells	8,803
Brown skins from		Pecan shells	
peanuts	10,431	(few meats left)	10,444
Corn on the cob	8,100	Coffee grounds	10,058

[a] Data from [6].

would be 0.6 lb and that this would be supplemented by about an equal weight of dry combustible material to make a total of 1.2 lb/(person) (day), or nearly 5 lb/day for an average family of four persons [7].

The laboratory test charge finally selected consisted of equal parts of food wastes and dry combustibles, as shown in Table 4A–3.

Table 4A–3   Analysis of Basic Load Used to Test Incinerating Ability

Constituents	Composition (percent)	Weight per Single Charge (ounces)	Particle Size
		Foodstuffs	
White potatoes	35	14.0	$\frac{1}{2}$-in. square slices
Cabbage	15	6.0	$\frac{3}{4}$-in. cubes
Water	13	5.2	(a)
Oranges	10	4.0	$\frac{1}{8}$-in. segments
White bread	10	4.0	Half slices
Beef suet	5	2.0	$\frac{3}{4}$-in. cubes
White rice	12	4.8	Whole grains
		Dry Combustibles	
Corrugated cardboard	50	20	6-in. squares
Newspaper	25	[b]10	Double-page sheets (crumpled)
Waxpaper	25	10	3-ft lengths (crumpled)

[a] Added as surface moisture.
[b] Newspaper used for wrapping food items is included in this weight.

Using a standard laboratory charge, two smokeless-odorless units were developed with a charging capacity of $1\frac{1}{2}$ bushels. The desired performance was obtained in the two prototype incinerators by the following measures:

1. Using combustion air and flue-gas-dilution air for cooling the jacket and flue gases.

2. Controlled combustion-air supply for initial burning of the charge, with separate combustion-air supply for afterburning.

3. Using a gas-fired afterburner for maintaining about 1500°F temperature and completely oxidizing smoke and odors.

4. Impingement or change of flow direction and velocity reduction to drop out fly ash.

5. A barometric damper for diluting and cooling effluent gases and preventing excessive chimney draft.

6. An ash-drawer capacity of at least 20 percent of the loading capacity.

7. Ready access to burners and controls for servicing.

The supplementary fuel required to burn domestic garbage satisfactorily was found to vary from 2655 to 8130 Btu per pound of charge, depending on the amount of paper in the charge. Total gas-input rates of the two prototype units developed were 40,000 to 42,000, most of which was necessary for thorough afterburning. Figure 4A–1 shows a cross section of one of the prototypes [7].

Following the example of the two prototype units, many incinerator manufacturers developed their own units incorporating the control and afterburning principles that have been demonstrated [7].

Operation of the new smokeless-odorless, afterburner incinerators is similar to that of the older high-Btu-input type; but in general they have a considerably greater heat input than the older types because the major portion of the gas flame is discharged into the secondary chamber for smoke and odor destruction. Some units have a timer on the main gas-burner line that keeps the burner on for at least 2 hours after it is ignited.

When the main gas burner is lighted the dry materials in the incinerator are ignited, and the temperature of the gases leaving the unit rises rapidly. This increase in temperature is often less rapid than in the old-type incinerator, and hence the secondary burner is relied on for smoke and odor reduction. The temperature in the secondary chamber should rise to a level of 1200 to 1500°F to control odors. Most of the dry paper and similar materials are destroyed quickly, and the remainder of the refuse burns slowly.

Even though the afterburner type of incinerator is used to dry and burn the refuse, it takes more time to consume a given amount of refuse than a high-Btu-input unit does. Two-thirds of the gas input is used to destroy smoke and odors. Additional fuel can be directed into the primary combustion chamber to speed up combustion, but that will increase the amount of gas consumed. If the burner is to be operated within the 1200 to 1500°F range, the unit must have a refractory lining adequate to withstand such temperatures.

A typical incinerator is shown in Figure 4A–2. The primary chamber is for incineration; the secondary chamber is for burning smoke and

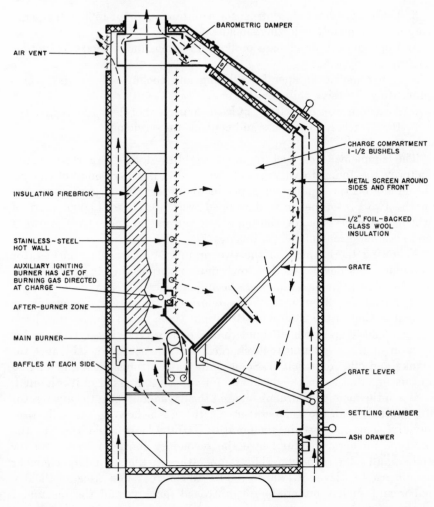

Figure 4A–1 Cross section of prototype smokeless-odorless, afterburning incinerator. Gas-input rate is 40,000 Btu/hr [7].

odorous constituents, and collecting fly ash [4]. A number of gas-burner arrangements are used by various manufacturers. These include the following:

1. A burner that fires horizontally through a tube into the primary chamber; in most designs, however, the burners continue to the back chamber for burning smoke and odorous constituents.

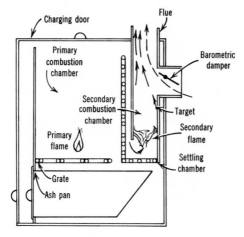

Figure 4A–2 Features of a typical smokeless-odorless domestic incinerator [4].

2. A horizontal tube without openings that extends through the primary chamber, with all heating and ignition done by radiation through the tube. This tube is bent upward in the secondary chamber, where a metal baffle above it diverts its heat and flame to mix with the smoke and odorous constituents for ignition and burning.

3. A burner across the back of the base of the unit, with part of the flames directed at an angle to the primary chamber for incineration, and the major portion directed upward into the secondary chamber for ignition and burning of smoke and odorous constituents.

4. A power burner that directs a flame at the top of the pile of garbage.

AIR-POLLUTION POTENTIAL

Table 4A–4 shows the air-pollution potential of the various types of domestic units compared with other sources of pollution [2]. Comparable data are not available to show the effects of age, use, and abuse of these devices. Some deterioration in performance is to be expected, depending on severity of use. The reliability of smokeless-odorless units in maintaining nuisance-free performance can be expected to increase as design, materials, and manufacturing techniques improve.

ECONOMICS OF ON-SITE DOMESTIC INCINERATION

Just as moving about by mass transportation is more economical and potentially cleaner than by individual cars, so central, municipal inciner-

Table 4A–4 Trace Effluents in Stack Emissions from Domestic Gas-Fired, Municipal, and Commercial Incinerators, and Backyard Trash Burners, Compared with Gas- and Oil-Fired Heating Units and Automobile Exhaust[a]

Source	Aldehydes[b]		Nitrogen Oxides[c]		Organic Acids[d]		Ammonia		Hydrocarbons[e]		Sulfur Oxides[f]		Carbon Monoxide (ppm)	Particulate Matter[g] (gr/ft^3)		Smoke Compliance with ASA Requirements[j]
	ppm	lb/ton	ppm	lb/ton	ppm	lb/ton	ppm	lb/ton	ppm	lb/ton	ppm	lb/ton		At Standard Temperature and Pressure[h]	At 500°F[i]	
Domestic gas-fired incinerators:																
AGA prototype:																
Shredded paper	8–21	0.9–2.3	6–13	1.0–2.2	—	—	<5	—	—	—	—	—	100	0.017–0.019	0.023–0.028	Yes
ASA domestic wastes	8	0.8	15	2.1	7	1.8	<5	—	0.7	0.3	2	—	200–400	0.005–0.018	0.006–0.026	Yes
New manufacturers' units:																
Shredded paper	4–67	0.7–15.9	2–7	0.3–2.6	—	—	<5	—	—	—	—	—	200–400	0.013–0.095	0.030–0.222	Yes
ASA domestic wastes	25–40	—	2–5	—	—	—	<5	—	—	—	—	—	200–1000	0.006–0.012	0.011–0.026	Yes
Older units:																
Shredded paper	24–48	0.3–0.4	5	0.6–0.8	17	6.6	5	—	4.7	2.5	—	—	—	0.039–0.132	0.084–0.282	No
ASA domestic wastes	5–30	5–6	1–3	0.6–2	—	—	5	—	—	—	—	—	—	0.019–0.097	0.122–0.526	No
Municipal incinerators:																
Glendale, Calif.:																
With scrubber	1–9	—	24–58	—	—	—	—	—	None	None	—	—	<1000	0.035–0.060	0.12–0.16	—
Without scrubber	1–22	—	58–92	—	—	—	—	—	None	None	—	—	<1000–3000	0.128–0.347	0.14–0.066	—
Other units:																
Single chamber	—	0.03–2.7	—	3.9–4.6	—	2.0–3.9	—	0.33–0.5	—	None	—	1.4–2.3	—	0.75	—	—
Backyard:																
Paper and trimmings[j]	760	29	<1.5	<0.1	—	—	65	1.8	—	—	—	—	5500–27,000	—	—	—
6-ft³ paper	49	2.1	7	0.5	18	1.5	4	0.1	—	—	34	1.2	—	—	—	—
Industrial heating units:																
Large, gas fired	49	2	215	14	30	2.5	0.6	—	—	—	4	0.3	—	—	—	—
Large, oil fired	61	2.4	390	26	365	30	0.6	—	—	—	750	60	—	—	—	—
Automobile exhaust:																
Accelerating	1369	41	4180	190	—	—	—	—	410	30	—	—	<100	—	—	—
Cruising	264	8	1606	75	—	—	—	—	354	28	—	—	4000	—	—	—

a Data from [3].
b As formaldehyde.
c As nitrogen oxide.
d As acetic acid.
e As hexane.
f As sulfur dioxide.
g Including tarry organic materials.
h Actual.
i 50 percent excess air.
j Battelle.

ation, when it is properly done, can be cheaper and potentially cleaner than on-site disposal. However, since most municipal incinerators are operated today without high-efficiency pollution-control equipment, they are much more significant sources of pollution than the smokeless-odorless units. Two features of on-site disposal—namely, a fuel cost for the new units of generally less than 10 cents per day and the convenience of immediate, clean, storage-free waste disposal are highly attractive to many homeowners, comparable to the time and effort-saving convenience of their cars.

REFERENCES

[1] *American Civil Engineers' Handbook*, 5th ed., "Refuse Collection and Disposal," Wiley, New York, 1942, pp. 1750–1772.

[2] Doland, J. J., and E. W. Steel, "Municipal Sanitation," *General Engineer's Handbook*, 2nd ed., McGraw-Hill, Inc., New York, 1940, pp. 656–740.

[3] Hein, G. M., and R. B. Engdahl, *A Study of Effluents from Domestic, Gas-Fired Incinerators*, AGA, New York, 1959.

[4] Houry, E., and H. W. Kain, "Principles of Design of Smokeless-Odorless Incinerators for Maximum Performance," *AGA Res. Bull.*, **93** (1962).

[5] Kaiser, E. R., "Refuse Composition and Flue-Gas Analyses from Municipal Incinerators," *Proc. ASME National Incinerator Conf.*, New York, May 1964, pp. 35–51.

[6] Kent, R. T., *Mechanical Engineers' Handbook*, 12th ed., "Combustion and Fuels. Straw, Paper, and Miscellaneous Waste Fuels," Wiley, New York, 1950, pp. 2–44.

[7] Skipworth, D. W., G. M. Hein, and H. W. Nelson, "Design of Domestic, Gas-Fired Incinerators for Elimination of Smoke, Odors, and Fly Ash," *AGA Res. Bull.*, **78** (1958).

[8] Vandaveer, F. E., "Gas Incinerator Characteristics: Acceptance, Servicing, and Load," *AGA Monthly*, **36**, No. 10, 33–34 (October 1954).

B. MULTIPLE DWELLINGS

Harold G. Meissner

Rapid changes during the past few years in the characteristics and quantity of refuse collected in large apartment houses have greatly increased the problem of disposal and in many cases indicated the desirability of on-site incineration. The bulk and heating value of refuse per capita have increased as more people use more paper and plastic materials. Apartment dwellers buy prepared and packaged foods, with less vegetable and meat waste. Such waste now makes up only 10 to 15 percent of the total refuse from an average apartment house. In some cities it is fed through garbage grinders to the city sewers.

The uptrend in packaging foods, liquids, and other merchandise in metal, glass, and plastic containers has added to the cost of composting. Such materials must be removed one way or another, preferably at the source. This increases the cost of collection.

The much greater volume of present-day refuse handicaps landfill operations, reducing their available life to a fraction of previous expectations and making the location of additional areas more difficult.

Changes in papermaking processes have reduced the salvage value of waste paper to one-third or less of the value it had a few years ago. Often it is incinerated or put in landfill.

REFUSE QUANTITIES

Tests made in 1958 by Kaiser et al. [1] on high-rise-apartment incinerators in New York City showed the total weight of refuse from a typical 500-

---

Consultant and Professional Engineer. Formerly Assistant Director of Engineering, Department of Air Pollution Control, New York City.

population dwelling to be 720 and 850 lb/day, depending on whether or not newspaper was included. With newspaper the average per capita was 1.7 lb/day; without newspaper, it was 1.44 lb/day. Newspapers generally were bundled and put out for salvage or pickup by city trucks.

The bulk density of this refuse, as burned in the incinerators, averaged 4.1 lb/ft³. The total volume was 207 ft³ with newspapers and 175 ft³ without. The newspapers were charged loose into the furnace. Storing this refuse before incineration required 52 standard, 4-ft³ trash cans per day.

The total amount of refuse produced in many cities has increased materially, but most of this increase has been from industrial and other sources. The per-capita refuse in the average apartment house has shown no significant increase.

Except for slum areas with abnormal occupancy, population per building corresponds roughly to the number of living rooms over 60 ft² in area. There are generally two persons per bedroom and four persons per apartment. When more accurate information is not available it is possible to estimate refuse-disposal facilities from the above data.

REFUSE CHARACTERISTICS

The refuse normally deposited in multiple-dwelling incinerators differs materially from that delivered to central municipal or industrial incinerators, as described in other chapters.

Large cartons or crates are not present. Unless they are broken or cut into small pieces, they cannot be charged through the relatively small hopper doors.

Garden trash, branches, coal or wood ash, and similar outside debris are of course very limited or entirely absent—as is oversize noncombustible waste.

Plastic materials are limited to wrappings or containers from food and other merchandise, and are unlikely to be present in concentrations large enough to cause smoke problems.

Bottles and cans form a larger proportion of the refuse and comprise 60 percent or more of the residue.

The moisture content of the refuse is low, coming mostly from the garbage or remains of fresh food products, which average only about 10 percent of the total refuse.

The average heating value of the refuse is assumed to be 6000 Btu/lb as compared with the typical 5000 Btu/lb for municipal refuse.

Analyses of typical apartment-house refuse are shown in Table 4B–1.

Table 4B–1   Analyses of Typical
Apartment-House Refuse

Proximate Analysis (percent)	
Moisture	10.0
Volatile matter	59.3
Fixed carbon	8.2
Ash and metals	22.5
Total	100.0

Ultimate Analysis (percent)	
Total carbon	49.8
Hydrogen	6.6
Oxygen	42.8
Nitrogen	0.6
Sulfur	0.2
Total	100.0

ALTERNATIVE DISPOSAL FACILITIES

Refuse disposal in the multiple dwellings is generally limited to one or more of the following methods or combinations thereof:

1. It may be picked up on the various floors by the porter and carried down to the basement in bags or containers.

2. It may be taken to the basement by the tenants and put in cans.

3. It may be deposited in chutes or flues through hopper doors on each floor to fall into suitable containers or directly into an incinerator in the basement.

4. Garbage and other grindable refuse may be put into disposal units or garbage grinders in the sink, from which it is flushed into the sewer.

5. Newspapers may be bundled and put out for pickup by salvage handlers or city trucks. Cartons, crates, and other bulky or salvageable refuse may be handled similarly.

In the basement the refuse may be collected loose in cans, containers, or bags for subsequent pickup by city or private trucks, or it may be compacted in one of several systems to reduce its bulk and storage-space requirement. Some salvaging may be done at this point.

The refuse may be macerated and pulped to reduce its volume. The excess water may then be squeezed out to facilitate subsequent removal.

The weight of the refuse is increased materially by this method. Flushing to sewers is frequently prohibited.

The refuse may be incinerated by means of one of the several systems discussed below.

Compacting the refuse at the building reduces the volume to be stored to one-half or one-third of its initial raw bulk, but it poses the problems of odors and fire hazard. The storage room must be cooled and fire-proofed. In one method the refuse is forced into metal containers, from which it is subsequently dumped into pickup trucks. Here the refuse may regain much of its original volume. In another method the refuse is compressed into paper bags, which add appreciable weight and which may have to be broken apart for subsequent incineration at the central plant. In a third method the refuse is compressed into special metal containers, which are picked up and hauled to the disposal facility. Returning them to the pickup point adds hauling cost. In all these cases there is not a reduction but a possible increase in the weight of refuse and only a temporary reduction in volume. The refuse is still putrescible and flammable if deposited into landfills, where it poses problems of spontaneous combustion or other fires, noxious-odor emission, and vermin.

Compacting refuse at the building offers some advantage in reduced storage-space requirements but no improvement in ultimate disposal. It may actually add to the cost of disposal, as noted above.

Emergencies such as bad weather, strikes of building or trucking employees, and other unavoidable contingencies may prevent removal of the refuse for prolonged periods, in which case storage of the residue would present much less severe problems than would storage of the refuse; for example, refuse storage requires from 5 to 15 times as many cans or containers, all of which must be handled at least twice. Storage rooms may have to be fireproofed—and refrigerated to prevent putrefaction, vermin infestation, and similar health hazards.

On-site incineration has therefore a number of advantages over the other disposal methods described. These include reducing the refuse in volume and weight to a fraction of the original and finally to a sterile, odorless residue that may be hauled directly to the dump area or, if necessary, stored in the building without fire or health hazards. The great saving in both pickup-truck capacity and in fill-area requirements when residue only has to be handled is shown in Table 4B–2. Landfill life can therefore be extended almost 15 times when residue is used.

Because of these advantages of on-site incineration, emphasis should be on solving its inherent problems, such as possible air pollution, smokeout, and improper operation by inexperienced personnel. These may be solved by using the latest designs and techniques now available.

Table 4B–2   Relative Weights and Volumes per Ton
of Refuse Deposited, and Landfill Requirements[a]

	Loose Refuse	Compacted Refuse	Residue
Weight (lb)	2000	2000	500
Volume (ft³)	488	163	49
Trash cans required	122	41	12

	Raw Refuse	Residue
Landfill requirements (ft³)	72	5

[a] Developed from data collected by Kaiser et al. [1].

TYPES OF MULTIPLE-DWELLING INCINERATORS

There are two general types of incinerators in common use, the major difference being the method of feeding the refuse from the various floors into the furnace.

Flue-Fed Incinerators

Flue-fed incinerators are those in which refuse is charged through hopper doors on each floor into a refractory flue whose bottom opens directly into the top of the furnace or combustion chamber. In the *single-flue* design (see Figure 4B–1) the combustion products flow upward through the same refractory flue to be discharged above the roof of the building, whereas in the *double-flue* design (Figure 4B–2) the combustion products normally pass upward through a parallel refractory flue. At intervals the hot gas from the auxiliary burner may be discharged through the charging flue to purge it of any odors or vermin that may have accumulated from the refuse rubbing against the walls. Both flues should be open on top to allow free exit of any combustion products.

Chute-Fed Incinerators

Chute-fed designs are those in which the refuse is charged through hopper doors on each floor into a metal flue, which may or may not be lined with refractory or other material. The refuse falls into a basement hopper from which it is transferred either manually or mechanically into the incinerator. Combustion products flow up through a refractory flue, to be discharged above the building's roof line. A typical example of a chute-fed incinerator is shown in Figure 4B–3.

PERFORMANCE VERSUS CONSTRUCTION CRITERIA

Satisfactory operation of new installations may be ensured by using one of several types of criteria or standards, including performance standards, construction specifications, or combinations of the two. In performance standards minimum requirements are prescribed for such things as unburned combustible matter in the residue, smoke, and fly-ash or dust loadings in the combustion products. In combustion specifications criteria are established for sizes or dimensions of furnaces, flues, materials for construction, and similar details.

Performance standards may be implemented with tests of prototypes of proposed designs or of each installation as it is installed. Since such tests are quite expensive and require considerable personnel, this procedure can become rather costly and involved when many new applications are processed.

Construction specifications are based on previous testing of the equipment and wide experience of engineers with design and operation of incinerators. Through flexibility in these criteria advantage can be taken of new developments in equipment and materials as they become applicable.

Suggestions for implementing performance standards and construction specifications are outlined below, but the ultimate responsibility for satisfactory performance to meet the local air-pollution-control code rests with the installer and user.

Performance Standards

Many localities permit installation of incinerators in accordance with the filer's plans and specifications but withhold final approval until completion of satisfactory performance tests of each installation or analysis of results of previously tested prototypes. Some facilities may accept manufacturer's or installer's guarantee in lieu of such tests, depending largely on the experience and reliability of the guarantor.

Dust-loading tests on incinerators require specialized knowledge and equipment, which are not always available. During tests the incinerator should be operated at its approved rating in pounds per hour, and the refuse used should be as typical as possible of that normally fired.

The cost of such tests will range from $500 to several thousand dollars, depending on the number of tests involved and local conditions, including

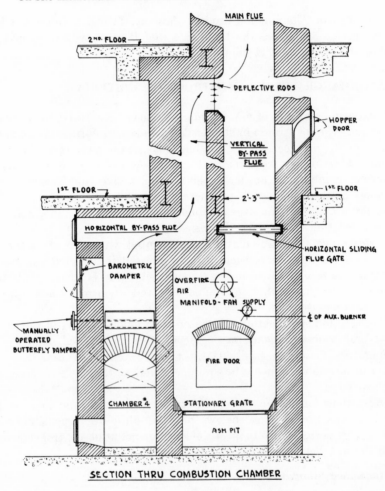

Figure 4B–1 Single-flue incinerator with fly-ash collection chambers and bypass flue.

the size and complexity of the flue and facilities for test probes. On large flues a number of traverse points are required at right angles to one another, especially when long, straight runs of flue or stack are unobtainable. In some cases straightening vanes may be required to avoid stratification and questionable gas sampling. In one form or another the cost of such tests would have to be included in the price of the incinerator. The burden of proof must be borne by the installer or user rather than by the enforcement agency.

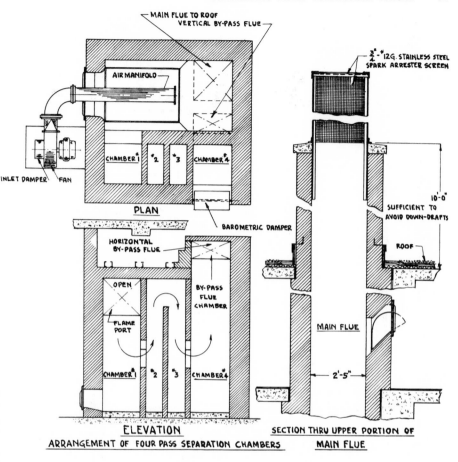

Figure 4B–1 Continued.

The eventual development of continuous-gas-sampling equipment, now under way, will greatly simplify the enforcement of performance standards, provided the equipment is maintained in satisfactory condition. Individual tests give results only for the conditions existing at the time of testing. These conditions may change materially with the type of refuse charged and other variables.

Construction Specifications

A typical example of these criteria is taken from specifications used for several years in New York City. It incorporates the latest available

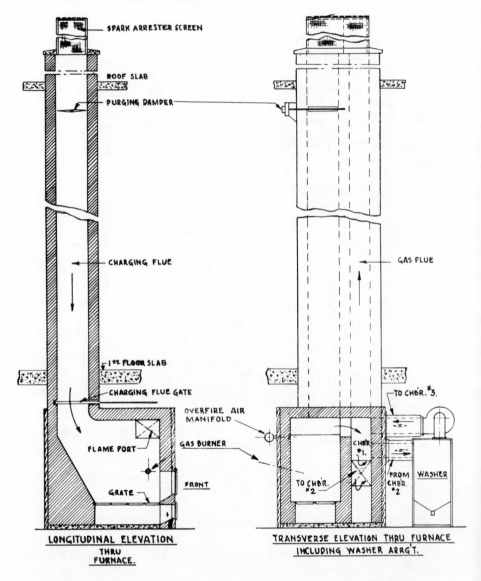

Figure 4B–2 Double-flue incinerator with gas washer and induced-draft fan.

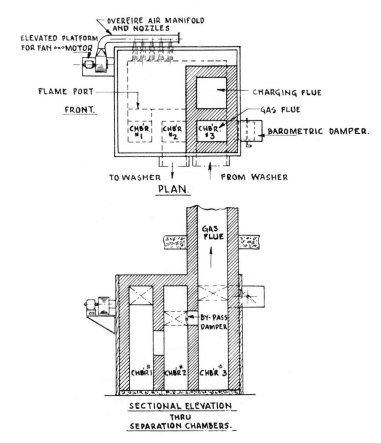

Figure 4B–2 Continued.

equipment and construction details. It requires filing of complete plans
and specifications, which are examined by a staff of competent engineers.
Any discrepancies found are detailed on an objection form, which is
sent to the filer. The filer then submits an amendment form covering
the various items listed.

When the examiner finally approves the application a work permit is
issued to the filer so that he may proceed with the installation. The
installed incinerator is inspected by an engineer to ensure that the work
has all been done satisfactorily and in accordance with the approved
work permit. When the incinerator is approved a certificate of operation
is issued, copies of which are kept on file at the building.

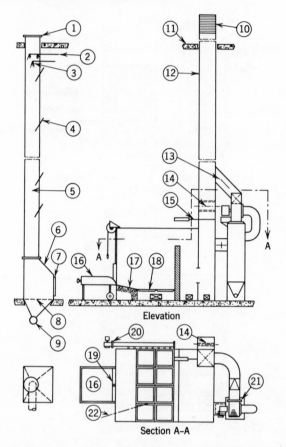

Figure 4B–3 Chute-fed incinerator and adjacent refuse chute. 1. Safety cap for fire protection. 2. Detergent spray for purging. 3. Sprinkler head for fire protection. 4. Hopper door. 5. Refuse chute. 6. Refuse hopper. 7. Cleanout door. 8. Grating for drainage. 9. Waste water to sewage. 10. Stainless-steel screen. 11. Roof slab. 12. Gasflue. 13. Draft-control damper. 14. Barometric damper. 15. Bypass damper (automatic control). 16. Charging apron. 17. Hearth. 18. Grate area. 19. Guillotine charging gate (power operated). 20. Overfire-air manifold and fan. 21. Washer and induced-draft fan. 22. Auxiliary burner.

Approval of the installation in accordance with the above procedure does not relieve the user of responsibility for satisfactory performance of the incinerator in accordance with the local air-pollution-control code.

Rather than being restrictive, well-developed construction criteria can benefit the manufacturers of adequate and well-constructed equipment by setting up minimum requirements that must be met and letting them know beforehand what conditions will be approved.

Such criteria are similar to the specifications issued by architects or engineers covering the equipment involved and are in accord with common codes developed through the years to set up minimum requirements and ensure satisfactory performance, regardless of varying designs and materials offered by the contractors. In the author's experience with incinerator proposals differences of as much as 2:1 have been noted in such vital factors as furnace sizes for the same job. This would obviously give an unfair advantage to manufacturers offering the smaller equipment, which is presumably less adequate.

Example of Construction Criteria

General. The incinerators shall comprise steel-cased, refractory-lined furnaces; equipment for fly-ash removal such as washers, scrubbers, precipitators, or other approved types; double flues extending through the roof of the building and provided with suitable spark screens; combustion-control equipment for automatic operation, including a 24-hr cycling clock; all as described below.

Furnace Parameters. Internal dimensions of the primary combustion chamber, henceforth called the furnace, shall be in accordance with the tabulation curves shown in Table 4B–3 which gives the minimum arch height, furnace length, projected area including grate and hearth, and volume, based on the number of rooms or population, whichever is greater. The heat-release rates from which these dimensions are calculated are also shown in this tabulation.

Combustion Air and Products. These are plotted in Figure 4B–4, which also includes the duct velocities from which minimum duct sizes shall be determined. Note that the gas volumes are given in the three values commonly experienced, which include (a) that leaving the furnace, (b) washed and cooled gas to the induced-draft fan, and (c) air-cooled gas to the flue when the washer is bypassed for some reason, such as a power failure. If the power should fail, the barometric damper would admit tempering air to prevent excessive flue temperatures.

Auxiliary Heat Source. To ensure quick ignition of the incoming refuse and adequate temperature for complete combustion, auxiliary fuel (such as gas or oil) shall be fired into the furnace under temperature control. The minimum Btus per hour are given in Table 4B–3 and the preferred burner location is illustrated in Figures 4B–1, 4B–2, and 4B–3. Elevation shall be high enough above the grate to avoid fouling by falling refuse, with the burner angled downward to ensure rapid ignition of the fuel bed.

Table 4B-3 Typical Furnace Parameters

	Number of Rooms or Population per Incinerator						
	100	200	300	400	600	800	1000
Refuse per day at 1.44 lb/room (lb)	144	288	432	576	865	1152	1440
Volume of refuse per day at 4.1 lb/ft³ (ft³)	34.1	70	105	141	211	282	351
Heat input per day at 6000 Btu/lb (Btu)	864 000	1,728,000	2,590,000	3,460,000	5,180,000	6,912,000	8,640,000
Refuse per burn at 25%[a] (lb)	36	72	108	144	216	288	360
Volume of refuse per burn at 4.1 lb/ft³ (ft³)	8.8	17.6	26.3	35.2	52.7	70.2	87.8
Heat input per burn at 6000 Btu/lb (Btu)	216,000	432,000	648,000	865,000	1,300,000	1,730,000	2,160,000
Projected area heat release [Btu/(ft²)(hr)]	10,800	20,000	29,000	35,000	48,400	57,667	65,000
Projected area—grate and hearth[b] (ft²)	20	22	23	24	26.9	30	33
Furnace length and width inside[b]	6'4" × 3'2"	6'7" × 3'3¼"	6'9¼" × 3'4¼"	7'0" × 3'6"	7'4" × 3'8"	7'7" × 3'9½"	8'2" × 4'1"
Grate area at 50% of burning area[b] (ft²)	10	11	11.5	12	13.5	15	16.5
Arch height above grate[b] (ft)	4'0"	4'3"	4'6"	4'9"	5'0"	5'4"	5'9"
Basement height under beams or slab[b] (ft)	9'8"	9'11"	10'2"	10'5"	10'8"	11'0"	11'5"
Furnace heat-release rate [Btu/(ft³)(hr)]	2800	4800	6270	7600	9680	10,500	11,400
Furnace volume (ft³)	80	90	103.5	114	134.5	160	190
Gas weight leaving furnace at 200% excess air (lb/hr)	440	880	1323	1760	2650	3520	4415
Gas volume leaving furnace at 1600°F (cfm)	379	758	1138	1515	2280	3030	3900
Gas volume in flue (after barometric damper) at 500°F (cfm)	685	1370	2055	2740	4120	5480	7050
Combustion-air weight (lb/hr)	405	810	1215	1620	2440	3240	4050
Combustion-air volume at 80°F (cfm)	90	180	276	367	554	735	920
Underfire air at 50% of total at 80°F (cfm)	45	90	138	183	277	367	460
Adjustable air-port area[b] (in.²)	5	10	15	20	31	42	51
Overfire-air fan at 25% of total at 1-in. static pressure (cfm)	22.5	45	69	92	139	184	230
Overfire-air-duct and manifold area at 2000 fpm (in.²)	1.62	3.24	4.97	6.62	10.0	13.25	16.5
Equivalent schedule 40 pipe size (in.)	2	2	3	3	4	4	6
Number of 1-in. pipe nozzles for overfire air	[b]4	4	5	5	6	7	9
Flame-port area at 1000 fpm (ft²)	[b]1.0	[b]1.0	1.5	1.5	2.3	3.0	3.9
Separation-chamber port area at 2000 fpm (ft²)	[b]1.0	[b]1.0	1.0	1.0	1.2	1.5	1.95
Gas flue—barometric damper and fresh-air inlet[b] (ft²)	[b]1.0	1.37	1.86	2.36	3.29	4.05	4.86
Auxiliary-burner capacity at 1200 Btu/lb refuse[b] (Btu)	43,200	86,400	129,700	173,000	259,000	345,000	432,000

[a] Four burns per day.
[b] Minimum.

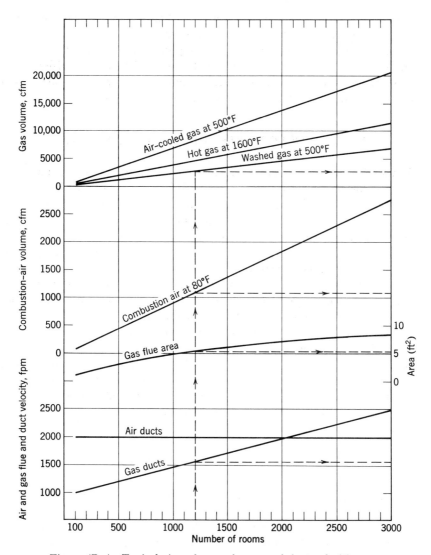

Figure 4B–4 Typical air and gas volumes, and duct velocities.

Auxiliary Burners. These shall preferably use natural gas and be
equipped with a spark-ignited gas pilot. They shall include flame failure
shutoff and furnace-draft interlock to prevent ignition when draft is
inadequate. All elements such as pilot burner, spark-electrode rod, and
temperature controller shall be readily accessible and in accordance with

Underwriters Laboratories requirements. The furnace temperature shall be maintained in the range of 1400 to 1600°F.

Automatic Control Equipment. This equipment shall be mounted in a suitable enclosed cabinet of steel plate. It shall include a 24-hr timing clock with relays to cause the various functions to occur in the required sequence and to provide fail-safe operation at all times.

COMBUSTION REQUIREMENTS FOR GOOD INCINERATION

Elimination of smoke, fly ash, odors, and other offensive emissions from incinerators must start in the furnace. The basic factors for accomplishing this include adequate time for combustion, adequate turbulence to ensure complete mixing of combustible matter and air, and the necessary temperature to complete the chemical reactions involved.

These are known as the "three Ts of combustion," and apply to incineration just as effectively as they do to the burning of any other fuel with special characteristics in size and handling; they are subject to the same requirements as other solid fuels, such as adequate combustion air, proper furnace design, and good operation.

The essential principles of combustion are fully covered in Chapter 2. Emphasis here is on incinerator design and operation. In general it is desirable to complete combustion of the solid and gaseous constituents in the furnace or primary chamber (where the maximum temperature occurs) to dry out the moisture and liberate volatile combustibles with the assistance of the auxiliary burner when necessary. The furnace should be adequate for these functions.

IMPLEMENTATION OF DESIGN REQUIREMENTS

Refuse Charging

Refuse is introduced through the hopper doors on the various floors at the convenience of the tenants. The amount varies widely from hour to hour and from day to day, depending on the type of occupancy. Business people, who are away all day, deposit some refuse after breakfast and again in the evening, whereas families with school children may make several deposits during the day. Uniform feeding of refuse to the furnace is therefore impossible as it is usually impractical to provide a storage hopper of sufficient capacity because of limited space. Incremental feeding of small charges may be accomplished by use of the charging-flue gate (see Figure 4B–1). Normally this gate should be closed, opening

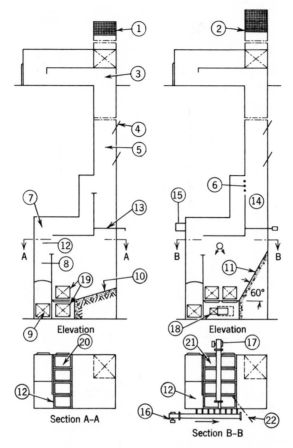

Figure 4B-5 Single-flue, multiple-chamber incinerator with bypass flue. Shown at left is an existing incinerator; at right, an upgraded design. 1. Low galvanized-wire screen. 2. High stainless-steel screen. 3. Roof settling chamber (optional). 4. Hopper door (locks optional). 5. Charging and gas flue. 6. Stainless-steel bars. 7. Bypass flue. 8. Settling chamber. 9. Cleanout door. 10. Flat hearth. 11. Steep hearth. 12. Flame port. 13. Charging-flue gate. 14. Charging-flue gate—power operated. 15. Barometric damper. 16 and 17. Overfire-air manifold and fan, and alternate. 18. Underfire-air register. 19. Fire and cleanout doors. 20. Inadequate grate area. 21. Enlarged grates. 22. Auxiliary burner.

automatically at 15- to 30-min intervals just long enough to let accumulated refuse drop into the furnace, except as noted hereafter. In this way any fumes or smoke generated in the furnace must pass upward through the gas flue and not cause odors or smokeouts through the hopper doors. Possible alternative methods for feeding the refuse in

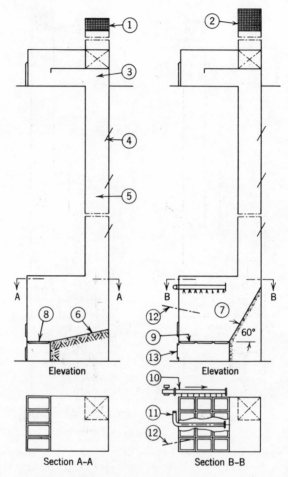

Figure 4B-6 Single-flue, single-chamber incinerator with roof settling chamber. Shown at left is an incinerator; at right, an upgraded design. 1. Low galvanized-wire screen. 2. High stainless-steel screen. 3. Roof settling chamber (optional). 4. Hopper door (locks optional). 5. Charging and gas flue. 6. Flat hearth. 7. Self-cleaning hearth. 8. Inadequate grate. 9. Enlarged grate. 10. Outside overfire-air manifold and fan. 11. Alternate inside manifold. 12. Auxiliary gas burner. 13. Underfire-air register.

small increments would include rotary, or star, feeders, which are in current use for handling other bulky materials, as well as pusher or plunger types, which force the refuse from the bottom of a hopper into the furnace with an alternate forward and rearward movement of a pusher or ram. Various other types of conveyors have also been tried.

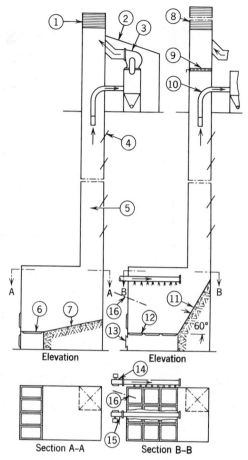

Figure 4B–7 Single-flue incinerator with washer or precipitator on roof. Shown at left is an existing incinerator; at right, an upgraded design. 1. Low galvanized-wire screen. 2. Washer enclosure. 3. Washer and induced-draft fan. 4. Hopper door (locks optional). 5. Charging and gas flue. 6. Inadequate grate area. 7. Flat hearth. 8. High stainless-steel screen. 9. Bypass damper with remote control. 10. Gas inlet to washer. 11. Steep hearth. 12. Enlarged grate area. 13. Underfire-air register. 14. Outside overfire-air manifold and fan. 15. Alternate inside manifold. 16. Auxiliary burner.

DESIGN REQUIREMENTS FOR GOOD INCINERATION

Refuse charging should be as nearly continuous as possible or at least in small increments rather than in large batches, so that momentary overloading is avoided. This may be accomplished by periodic opening of the charging-flue gate, say at 15- to 30-min intervals, or by using a rotating star feeder or similar means.

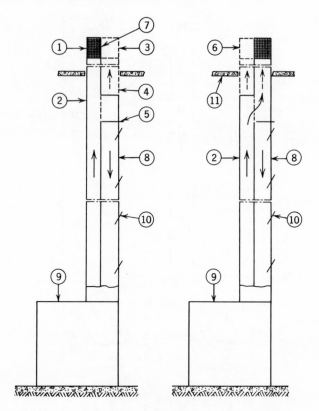

Figure 4B–8 Double-flue incinerator with one flue only extending through roof.
1. Partial screen. 2. Gas flue. 3. Enlarged screen. 4. Extended charging flue. 5. Purging
damper. 6. Extended gas flue and screen. 7. Partition screen. 8. Charging flue.
9. Incinerator. 10. Hopper door. 11. Roof slab.

An auxiliary burner or burners should be installed above the refuse in
the furnace to ignite the refuse and preheat the furnace as well as to
establish the draft at the beginning of each burning cycle, with temperature control for both maximum and minimum requirements.

Overfire and underfire air should be introduced by means of suitable
fans to ensure adequate combustion air and furnace turbulence.

Furnace configuration should be such as to provide the required grate
surface under the fuel bed, optimum arch height for adequate flame travel,
and optimum retention time for the combustion products.

Fly-ash- or particulate-removal facilities must include settling or baffle
chambers, washers, scrubbers, or other equipment to reduce dust loading

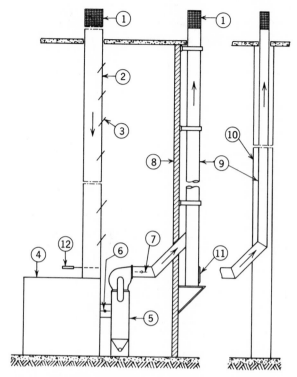

Figure 4B-9 Conversion from single-flue to double-flue incinerator. 1. Stainless-steel spark screen. 2. Existing flue. 3. Hopper door. 4. Incinerator. 5. Scrubber and induced-draft fan. 6. Tight damper. 7. Gas-flue-control damper. 8. Building wall. 9. Added gas flues. 10. Existing shaft or duct. 11. Cleanout door. 12. Charging-flue gate.

to acceptable limits, as well as suitable induced-draft fans to overcome draft loss when the stack draft is insufficient for this purpose.

Furnace draft equipment must include barometric dampers or other means for maintaining a low, uniform draft in the furnace at all times.

UPGRADING OF EXISTING INCINERATORS

The performance of existing incinerators may be greatly improved by revamping the furnaces, and installing controls and equipment in accordance with the design factors described above. The layouts shown in Figures 4B–5 through 4B–10 illustrate the designs in common use and what can be done to upgrade them with minimum cost and fewest changes in the surrounding structures.

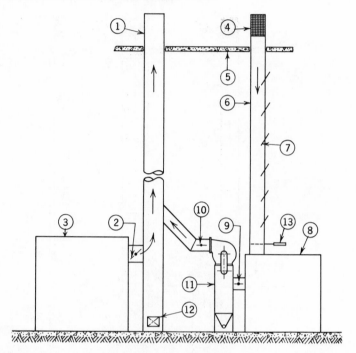

Figure 4B–10 Use of boiler chimney for incinerator gas flue. 1. Boiler chimney, no screen. 2. Draft-control damper. 3. Boiler with controls. 4. Stainless-steel spark screen on incinerator flue only. 5. Roof slab. 6. Charging flue for refuse and emergency gas flue. 7. Hopper doors in existing gas flue. 8. Incinerator. 9. Tight shutoff damper. 10. Gas-flow control. 11. Washer, scrubber, etc. 12. Cleanout door. 13. Charging-flue gate.

The minimum requirements for such upgrading should include the following items:

1. Incinerator capacity must be adequate for the refuse load based on a minimum of four 1-hr burning periods per day. These periods should be spaced to cover the refuse-load intervals.

2. Grates, brickwork, and other essential parts shall be in good condition. Hearth shall be steeply pitched rather than flat so that refuse will slide down onto the grate for complete combustion.

3. Auxiliary-burner capacity shall be in accordance with criteria given in Table 4B–3 to provide ignition of the incoming refuse and ensure adequate furnace temperature for complete combustion.

4. Overfire-air fan with manifold and nozzles shall be adequate to produce turbulence and combustion air for optimum burnout of both gas and carbon.

5. Automatic-control equipment, including programming clock, shall all be enclosed in a suitable steel cabinet.

6. Fly-ash-removal equipment, as noted above, shall be adequate to meet local air-pollution-control ordinances.

7. The incinerator room shall have a fresh-air inlet adequate for combustion-air requirements.

REFERENCE

[1] E. R. Kaiser, J. Halitsky, M. B. Jacobs, and L. C. McCabe, *Performance of a Flue-Fed Incinerator*, Technical Report 552.1, Research Division, College of Engineering, New York University, March 1958.

5

ON-SITE INCINERATION OF COMMERCIAL AND INDUSTRIAL WASTES WITH MULTIPLE-CHAMBER INCINERATORS

Ralph E. George and John E. Willamson

Disposal of combustible wastes continues to be one of the most perplexing problems in a modern urban society. All elements of the urban community—including industry, commerce, and the public at large—contribute to this problem. Of the several possible alternatives available for the disposal of combustible wastes, landfill disposal and incineration have been the most widely used and have received the most thorough consideration and investigation. In some large urban areas, such as the city of Los Angeles and its environs, with its complications of meteorology and population, many significant problems are presented by both of these alternatives. Sanitary landfill disposal, perhaps the least of two

Senior Air-Pollution Analyst and Senior Engineer, respectively, Los Angeles Air-Pollution-Control District, Los Angeles, California.

evils insofar as its contribution to the air-pollution problem there is concerned, requires the existence of suitable land sites within the range of economic hauling. With the expansion of landfill disposal in a densely populated and widespread area such as this conveniently available sites are rapidly depleted, and only remote landfill sites remain to be utilized.

In recent years increasing attention has been given to incineration as a medium for combustible-refuse disposal, not only to provide a suitable alternative in itself but also to complement the use of landfills by assisting in reducing overall volumes and conserving landfill areas.

However, to be considered as a suitable alternative, refuse disposal by incineration in any urban community must be accomplished without seriously contributing to atmospheric pollution. Although refuse disposal by incineration has been widely studied in the past, primary emphasis was placed on economic considerations. Interest in the air-pollution aspects was seemingly slight, and responsibility for design development was not generally assumed by the incinerator industry. As a result no basic advancements in incinerator designs were made for many years.

Early in 1949 the Air-Pollution-Control District of Los Angeles County recognized the continuing need for incineration as a medium for refuse disposal, as well as its potential as a contributing factor to the air-pollution problem. As a consequence the District undertook a program of investigation and tests to correlate the relationships of incinerator-design criteria to their effect on performance and on the discharge of combustion contaminants. To meet the needs of immediate problems initial investigations were confined to the single-chamber categories of incinerators. Shortly, these were proved culprits, and action was taken to condemn their use in Los Angeles County. Subsequent investigations were extended to the development of satisfactory multiple-chamber incinerator designs [1].

In many areas of the country the burning of combustible wastes in single-chamber incinerators, as well as in open fires, is still tolerated. A necessary prerequisite to the prohibition of such practices would ordinarily include the availability of satisfactory alternatives; for example, adequate municipal collection services and incineration in properly designed multiple-chamber incinerators. In some areas there is the problem in municipal collection that commercial and industrial establishments, as well as the majority of large multiple dwellings, are not provided with collection or else are not adequately served by municipal collection. Many establishments in these categories are often faced with circumstances that allow no alternative other than incineration for the disposal of their refuse. These may be such fundamental considerations as lack of storage space;

inaccessibility of storage space; the crippling effects of irregular, infrequent, or inadequate collection; or often the mandatory requirements for incineration of contaminated wastes. Another very realistic and important consideration is that of economic factors, which may dictate the necessity for incineration in lieu of other means of disposal.

The need for a satisfactory means of refuse disposal at the source resulted in investigations by the Los Angeles County Air-Pollution-Control District of incineration in all of its aspects. The object was to develop an efficient combustion furnace that would provide maximum reduction in waste bulk with minimum emission of air contaminants and would be capable of stable operation over a wide range of fuel mixtures and operating conditions. These studies have shown that the multiple-chamber incinerator offers the best potential for satisfying the requirements of complete combustion of refuse and operation within prescribed limits on contaminant discharge. Optimum operation was found to be dependent almost entirely on basic furnace design and the use of adequate design parameters.

Although an additional reduction in stack emissions may be obtained with the installation of properly designed wet collectors, elaborate and expensive control devices are not necessarily a prerequisite for satisfactory performance.

With the growing need for incinerators that are capable of complying with air-pollution-control codes and restrictions, it has become necessary to provide design data that will help make satisfactory incinerators available. Recommendations and standards are offered here to assist air-pollution-control officials faced with incineration problems, architects and engineering designers who must provide adequate designs, and manufacturers or contractors who will design as well as build incinerators that must meet air-pollution-control regulations and other standards of performance. Adequacy of design, proper methods of construction, and quality of materials are all important to the satisfactory completion of an incinerator that will meet air-pollution-control requirements and provide assurance of a reasonable service-life expectancy.

The design standards presented here provide the criteria for developing designs for multiple-chamber incinerators that can be expected to burn refuse with a minimum discharge of air contaminants. Tabular presentations alone are not sufficient for the best application and understanding of the principles and philosophies of design involved. It is also essential to understand the many factors that created the need for a new approach to incineration and the development of the multiple-chamber incinerator. The design recommendations and supplementary discussions offered here provide answers to many questions that confront designers and operators of multiple-chamber incineration equipment.

CLASSIFICATION OF REFUSE

The development of incinerator designs quite naturally begins with a basic knowledge of the types and quantities of refuse to be burned. It is particularly helpful to have information on certain specific properties, including chemical composition and moisture content. If this information is not available, the incinerator combustion calculations may be made, without introducing serious error, by considering the refuse to react stoichiometrically as cellulose (see Appendix 5–I).

A properly designed multiple-chamber incinerator can be expected to satisfy all incineration requirements, particularly when the material burned consists principally of dry refuse. It is not unusual, however, to find wide variations in moisture content, heating value, and other physical characteristics of refuse, and these can present some problems for the incinerator designer. It is extremely important that these factors be considered and that adequate provisions be incorporated into the design of a unit to accommodate such variations in refuse composition.

Generally refuse has a wide range of component proportions and an equally diverse range of physical characteristics. In the ensuing discussions, and for the purpose of the examples of design criteria and calculations given here, general refuse is considered as having component proportions within approximate limits, as follows:

Type 1 refuse, which consists of combustible wastes such as paper, cartons, rags, wood scraps, sawdust, foliage, and floor sweepings containing up to 25 percent moisture and up to 10 percent incombustible solids—and has a heating value of 6500 Btu/lb as fired.

Type 2 refuse, which consists of approximately an even mixture of refuse and garbage by weight, containing up to 50 percent moisture, 7 percent incombustible solids, and has a heating value of 4300 Btu/lb as fired.

INCINERATION

The empirical design relationships evolved from incinerator investigations and tests have resulted in the development of two basic types of multiple-chamber incinerators; namely, the retort type and the in-line type of multiple-chamber incinerator [4]. Other incinerator configurations—including incinerators with vertically arranged chambers, L-shaped units, and units with separated chambers breeched together—have appeared as variations of these two basic designs. Each style or configuration has certain characteristics with respect to construction and operation that limit its application.

Figure 5–1 shows a cutaway of a typical retort type of multiple-chamber incinerator. This type of unit derives its name from the return flow of effluent through the U-shaped gas path and the side-by-side arrangement of component chambers. Figure 5–2 shows a typical in-line design, so called because the various chambers follow one another in a line.

In both types of multiple-chamber incinerators the combustion process proceeds in two stages; that is, primary, or solid-phase, combustion in the ignition chamber, followed by secondary, or gaseous-phase, combustion. The secondary combustion zone is composed of two parts—a downdraft, or mixing chamber, and an up-pass expansion, or final combustion chamber. The gas flow and combustion reactions in the two-stage process proceeds as follows:

The ignition-chamber reaction includes the drying, ignition, and combustion of the solid refuse. As the burning proceeds the moisture and volatile components of

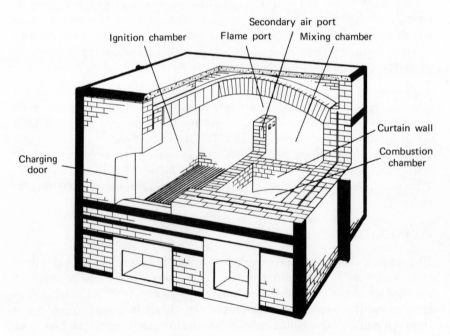

Figure 5–1 Cutaway of a retort-type multiple-chamber incinerator.

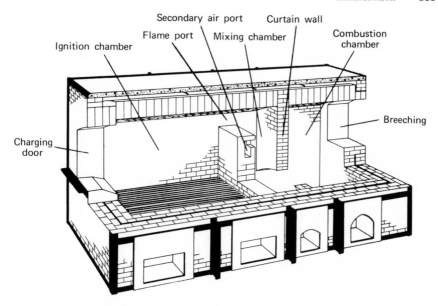

Figure 5–2 Cutaway of an in-line multiple-chamber incinerator.

the fuel are vaporized and partially oxidized in passing from the ignition chamber into the mixing chamber.

From the flame port the products of combustion and the volatile components of the refuse flow through the mixing chamber, at which point secondary air is induced. The combination of elevated temperatures and the addition of combustion air, augmented by mixing-chamber, or secondary, burners as necessary, assist in promoting the second stage of the combustion process. Turbulent mixing, resulting from restricted-flow areas and abrupt changes in flow direction, furthers the gaseous-phase reaction. In passing through the curtain-wall port from the mixing chamber to the final combustion chamber the gases undergo additional changes in direction, accompanied by expansion and final oxidation of combustible components. Fly ash and other solid particulate matter are collected in the combustion chamber by wall impingement and simple settling.

The gases finally discharge through a stack or, in some installations, through a combination of a gas cooler (e.g., a water-spray chamber or scrubber) and induced-draft system. Either draft system must limit combustion air to the quantity required at the nominal rated capacity of the incinerator.

Combustion Principles

Because of the heterogeneous nature of materials found in refuse, many factors cannot be predicted accurately except on an empirical basis. Theoretical treatment of the complex reactions taking place in some combustion processes is, as yet, incomplete, but the empirical art of combustion engineering has developed to an advanced state. The principles of solid-fuel combustion that in general apply to incineration and the basic precepts for combustion efficiency include the following:

1. Air and fuel must be in proper proportion.
2. Air and fuel, especially combustible gases, must be mixed adequately.
3. Temperatures must be sufficient for ignition of both the solid fuel and the gaseous components.
4. Furnace volumes must be large enough to provide the retention time needed for complete combustion.
5. Furnace proportions must be such that ignition temperatures are maintained and fly-ash entrainment is minimized.

The wide diversity in the chemical and physical properties of waste materials and the fluctuation in the charge in batch operations makes the development of satisfactory incinerator designs particularly challenging. In most other kinds of combustion processes fuel quality and fuel rates are predictable and usually are precisely controlled. This ideal situation does not prevail in most refuse-incineration processes. In addition to the wide variance in refuse composition, wetness, and volatility, there are differences in ash content, bulk density, heat of combustion, burning rate, and component-particle size with which to contend. All of these factors affect to some degree the operating variables of flame-propagation rate, flame travel, combustion temperatures, combustion-air requirements, and the need for auxiliary heat.

Comprehensive studies of operational characteristics and quantitative analysis of stack emissions from multiple-chamber incinerators have now made it possible to formulate empirical design criteria and to establish standards for evaluating the relationship of design parameters as they affect the discharge of combustion contaminants.

The important basic relationships presented here are those of grate loadings to combustion rates, arch heights to grate areas, length-to-width ratio of primary combustion chambers, flame ports, mixing chambers, secondary combustion chambers, and combustion-air supply and draft requirements. Proper evaluation and application of these design relationships will result in incinerator designs that are capable of operating at maximum burning rates with a minimum discharge of combustion contaminants (see Appendix 5–II).

Ignition Chamber

The ignition mechanism in a refuse-incineration process must be basically one of fuel-bed surface combustion. This is achieved by the predominant use of overfire combustion air and minimum use of underfire air. In this context overfire air is combustion air admitted into the ignition chamber at some point above the pile of refuse. Such air is generally furnished through air-supply ports located in, or adjacent to, the charging door on the front wall of the incinerator. Underfire air is combustion air introduced into the ashpit beneath the fuel bed through air ports located on, or adjacent to, the ashpit-cleanout doors.

By restricting the introduction of underfire air, relatively low fuel-bed temperatures are maintained, and entrainment of solid particulate matter in the effluent is minimized. If fuel-bed surface combustion through the use of overfire air is to be accomplished, the charging door must be located on the front wall of the ignition chamber or at the end of the chamber farthest from the flame port, because this is where the fresh charge of refuse is introduced. This method of introducing overfire air results in a movement of combustion air concurrent with the travel of the effluent, which has proved to be desirable for efficient combustion. With this design and method of charging the volatiles from the fresh charge pass through the stabilized, high-temperature zone above the fuel bed. In this way the rate of ignition of the refuse is controlled, preventing flash volatilization, flame quenching, and production of smoke, usually encountered when top- and side-charging methods are employed. The latter methods are generally unsatisfactory because they promote the suspension and entrainment in the effluent of dirt and dust mixed with the refuse, cause disturbance of the stabilized fuel bed, and require excessive stoking. With good control of the burning rate through proper charging of refuse and correct adjustment of combustion-air supply ports the need for stoking is reduced to only that necessary to move the fuel bed forward prior to introducing a fresh charge. Control of the combustion reactions and reduction in the amount of mechanically entrained ash and particulates in the ignition chamber are extremely important considerations in the design of an efficient multiple-chamber incinerator.

The emission of solid and liquid particulate materials in combustion effluents from multiple-chamber incinerators is principally a function of the mechanical and chemical processes taking place in the ignition chamber. The fundamental relationships to be considered in evaluating primary-combustion-chamber parameters are length-to-width ratio, arch height, and grate loading. Formulas governing ignition-chamber design are postulated from data obtained from tests of units of varying proportions operated at maximum combustion rates.

Length-to-Width Ratios

In the retort type of multiple-chamber incinerator, with rated design capacities of up to 500 lb/hr, very satisfactory operating results have been obtained with length-to-width ratios varying from 2.0:1 to 2.5:1. In units with design capacities of over 500 lb/hr optimum results are obtained with a length-to-width ratio of 1.75:1.

Although no fine line of demarcation has been established, optimum performance for burning rates ranging from 25 to 750 lb/hr has been obtained with the retort type of incinerator. Above this capacity it is difficult to obtain desirable combustion characteristics, proper flame travel, and combustion-air distribution and still retain the correct relationship of other critical design parameters. For burning rates in excess of 750 lb/hr the in-line type of incinerator is generally most satisfactory.

In the in-line type of multiple-chamber incinerator optimum length-to-width ratios commence at 1.65:1 for the 750-lb/hr capacity units and diminish linearly to 1.2:1 for incinerators with design capacities of 2000 lb/hr or more. Aside from the relationship of this parameter to combustion characteristics and the minimization of contaminant discharge, practical consideration should be given to the permissible ignition-chamber depth in manually stoked units.

Arch Height

Arch height has been observed to have an appreciable effect on contaminant discharge. Incinerators burning similar refuse and with similar grate areas but with different arch heights have varying combustion rates and contaminant-discharge characteristics.

Grate Loading

Acceptable grate loadings range from 15 to 25 $lb/(ft^2)(hr)$ for incinerators with burning rates from 25 to 750 lb/hr. Practical considerations for charging and stoking in the smaller incinerator units usually result in proportionately larger grate areas and lower grate loadings. For burning rates in excess of 750 lb/hr acceptable grate loadings range from 25 to 35 $lb/(ft^2)(hr)$. This upper limit may be exceeded in some instances by perhaps 10 to 15 percent, depending largely on the nature and composition of the refuse. A number of refuse incinerators have been designed to operate with grate loadings of 50 to 70 $lb/(ft^2)(hr)$. However, this has been accomplished by the use of underfire air as the predominant source of combustion air, the effect of which has been a marked increase in contaminant discharge.

Optimum values of arch heights and grate areas may be determined from Figures 5–3 and 5–4, respectively, by using the gross heating value of the refuse to be burned and hourly burning rates. The curves shown in each figure have an upper gross heating value of 9000 Btu/lb or more

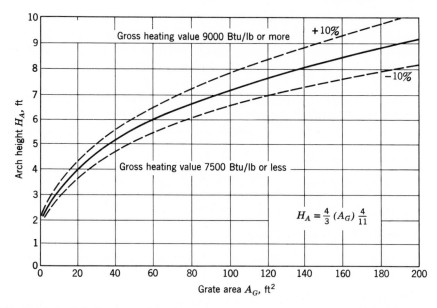

Figure 5–3 Relationship of arch height to grate area for multiple-chamber incinerators.

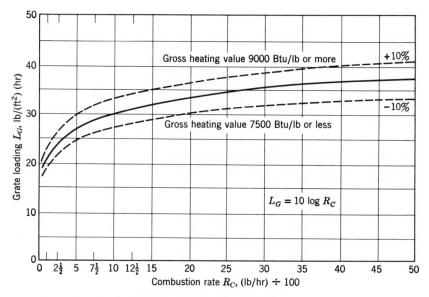

Figure 5–4 Relationship of grate loading to combustion rate for multiple-chamber incinerators.

and a lower gross heating value of 7500 Btu/lb or less. Interpolation between the upper and lower curve gives the correct arch height and grate area for refuse with a gross heat between these values. An allowable deviation from these values of plus or minus 10 percent is considered to be reasonable.

The relationship of arch height to grate area can be calculated from the following empirical equation for the average values of the two curves shown (this average curve corresponds to a heating value of 8250 Btu/lb):

$$H_A = \tfrac{4}{3}(A_G)^{5/11},$$

where H_A, arch height, is the average distance between the top of the grate to the arch in feet and A_G is the grate area or horizontal cross-sectional area of the ignition chamber in square feet.

Secondary Combustion Chambers

The gas-phase, or secondary, combustion reactions in a multiple-chamber incinerator are controlled largely by the flame port, mixing chamber, and secondary combustion chamber. The relationship of these parameters is usually determined by certain limiting gas velocities or by the unit volume requirements for the type and quantity of refuse burned. This entails a determination of the combustion-air requirements, weight and volume of combustion products to be handled, as well as combustion temperatures, heating values, moisture content, percent of combustibles, and ash content. These determinations and an evaluation of the combustion processes are usually made from an ultimate analysis of the refuse burned. However, this presents quite a problem when the refuse consists of a mixture of materials that vary widely in composition. The types of combustible refuse obtained from domestic, commercial, and industrial sources generally consist of such organic materials as grass, tree trimmings, wood, paper, and rags, which for the most part are chemically composed of cellulose. In view of this, unless a more accurate determination is possible, the combustion analyses may be made by correcting for moisture content and assuming the refuse to react stoichiometrically as cellulose $(C_6H_{10}O_5)$.

The factors controlling the design of the gas-phase combustion zone have been developed by applying fundamental evaluation precepts combined with experimentation on various proportions in the secondary chamber and port dimensions and on secondary air admission. It has been possible to establish from these investigations parameters for the secondary mixing and combustion-chamber portions of the multiple-chamber incinerator and to evaluate their relationship to gas velocities, turbulence, flame travel, degree of combustion, and contaminant discharge [1, 4]. The primary effect of proper design of the secondary mixing and combustion chambers and ports has been improved combustion of volatile and solid components. The last chamber, or final-combustion chamber, is intended to serve a dual purpose—it allows completion of the gas-phase combustion and also serves as a fly-ash settling chamber.

The flame port is designed to provide a high-gas-velocity zone. Gases leaving the flame port make an abrupt change in direction and are then expanded in the mixing chamber. The purpose is to promote turbulence and mixing of the effluent from the ignition chamber with secondary combustion air in a high-temperature-flame zone. Gas-phase combustion should essentially be completed in the mixing chamber. The cross-sectional area of the curtain-wall port separating the mixing chamber from the combustion chamber should be about 50 percent greater than that of the mixing chamber. This minimizes draft loss and prevents the effluent from sweeping the floor of the combustion chamber and entraining fly ash that may collect there. If the curtain-wall port is oversized, the effective length of the mixing chamber is reduced, which has a deleterious effect on the degree of completion of combustion of gaseous components.

Optimum design velocities determined for gas flows in the secondary combustion chambers and ports are summarized as follows:

Flame Port. Gas velocities at the flame port, or area provided above the bridge wall, should range from 45 to 65 ft/sec.

Mixing Chamber. Gas velocities in the mixing chamber may range from 20 to 35 ft/sec.

Curtain-Wall Port. The gas passage beneath the curtain-wall port should be designed to give gas velocities ranging from 10 to 15 ft/sec.

Secondary Combustion Chamber. Gas velocities in the secondary combustion chamber should not exceed 10 ft/sec.

Combustion Air

The final relationships to be considered in evaluating multiple-chamber design parameters affecting combustion efficiency and contaminant discharge are those of combustion-air supply and distribution, and the requirements for burners to supply auxiliary heat.

Essentially, enough air must be supplied to the incinerator to allow for the maximum combustion of the oxidizable materials charged; it must be introduced and distributed in such a manner as to reduce to a minimum the discharge of contaminants. This can be accomplished by providing air-port areas to allow for the introduction of at least 100 percent in excess of the amount of air theoretically required for complete combustion of the refuse. The air supply should be distributed so that not more than

10 percent of the total volume of air is introduced underfire or beneath the fuel bed. Approximately 70 percent of the air should be introduced overfire or above the fuel bed and 20 percent admitted through the secondary air ports into the mixing chamber. Where excess air is required to control incinerator temperatures the admission of this air should be made through additional overfire or secondary air ports. Secondary combustion air should be provided through controllable air-inlet ports located adjacent to the bridge wall. The air is introduced to the mixing chamber through openings in the bridge wall, just beneath the flame port.

Under normal operating conditions, at rated design capacity, primary combustion air supplied to the ignition chamber is sufficient to complete the combustion of volatile and solid combustible-effluent components reaching the mixing chamber. At times, particularly with highly volatile materials or when the charging rate is in excess of design capacity, there will be a deficiency of combustion air. On such occasions smoke and other combustible-effluent products pass through the flame port and into the mixing chamber. When this occurs it is essential that secondary air be supplied to the mixing chamber in order to complete combustion and provide smokeless operation of the incinerator. The oversize factors for the air-inlet ports, shown in Table 5–1, should be used to provide for these contingencies. In some incinerator applications there is a need for additional combustion air as well as supplemental cooling air over and above that calculated to be supplied by the air ports. This is particularly true in incinerators with uniform feed systems burning wood sawdust and shavings or in units burning other types of highly volatile refuse. In these special cases provisions for additional air must be made not only to accommodate the requirements of the combustion process but also to protect the refractory walls from damage that can result from excessive temperatures. This may require the doubling of air-port areas or the installation of supplementary air ports.

It is not unusual for multiple-chamber incinerators to operate with from 100 to 300 percent excess combustion air. Air-port areas ordinarily are sized to deliver about 50 percent of the total air required in the combustion process; that is, theoretical air plus about 100 percent excess air. The balance of excess air enters as leakage through expansion joints, through the charging door when refuse is introduced, and at other points of air leakage.

Overfire-air ports should be installed on the charging door located on the front wall of the incinerator. Any supplementary air ports should also be located on the front wall of the incinerator, at the end of the ignition chamber farthest from the flame port. Underfire-air ports should be located beneath the grates, and preferably at the same end of the

Table 5-1 Multiple-Chamber Incinerator Design Factors

Item and Symbol	Recommended Value	Allowable Deviation
Primary combustion zone:		
Grate loading, L_G	$10 \log R_C$ [lb/(hr)(ft^2)]; R_C equals the refuse-combustion rate in lb/hr	±10%
Grate area, A_G	$R_C \div L_G$ (ft^2)	±10%
Average arch height, H_A	$\frac{3}{4}(A_G)^{3/11}$ (ft)	—
Length-to-width ratio (approximate):	Up to 500 lb/hr, 2.5:1 to 2:1. Over 500 lb/hr, 1.75:1, diminishing from about 1.7:1 for 750 lb/hr to about 1.2:1 for 2000-lb/hr capacity. Oversquare acceptable in units of	
Retort		—
In line	more than 11 ft ignition-chamber length.	—
Secondary combustion zone:		
Gas velocities:		
Flame port at 1000°F, V_{FP}	55 ft/sec	±20%
Mixing chamber at 1000°F, V_{MC}	25 ft/sec	±20%
Curtain-wall port at 950°F, V_{CWP}	About 0.7 of mixing-chamber velocity	—
Combustion chamber at 900°F, V_{CC}	5 to 6 ft/sec; always less than 10 ft/sec	—
Mixing-chamber downpass length, L_{MC}, from top of ignition-chamber arch to top of curtain-wall port	Average arch height (ft)	±20%
Length-to-width ratios of flow cross sections:		
Retort, mixing chamber, and combustion chamber	Range—1.3:1 to 1.5:1	—
In line	Fixed by gas velocities due to constant incinerator width	—
Combustion air:		
Air requirement batch-charging operation	Basis: 300% excess air; 50% air requirement admitted through adjustable ports; 50% air requirement met by open charge door and leakage	
Combustion-air distribution:		
Overfire-air ports	70% of total air required	—
Underfire-air ports	10% of total air required	—
Mixing-chamber air ports	20% of total air required	—
Port sizing, nominal inlet-velocity pressure	0.1-in. water gage	
Air-inlet ports oversize factors:		
Primary air inlet	1.2	
Underfire-air inlet	1.5 for over 500 lb/hr to 2.5 for 50 lb/hr	
Secondary air inlet	2.0 for over 500 lb/hr to 5.0 for 50 lb/hr	
Furnace temperature (average temperature, combustion products)	1000°F	±20°F
Auxiliary burners (normal-duty requirements):	3000 to 10,000	
Primary burner	4000 to 12,000 Btu per lb of moisture in the refuse	
Secondary burner	$\begin{cases} 0.15 \text{ for } 50 \text{ lb/hr} \\ 0.30 \text{ for } 1000 \text{ lb/hr, uniformly increasing between sizes} \end{cases}$	—
Draft requirements:		
Theoretical stack draft, D_T	0.35 for 2000 lb/hr	
Available primary-air-induction draft, D_A (assume equivalent to inlet-velocity pressure)	0.1-in. water gage	
Natural-draft stack velocity, V_S	Less than 30 ft/sec at 900°F	—

119

incinerator as the overfire-air ports and charging door. Secondary air ports, already indicated, are located on the exterior wall of the incinerator adjacent to where the bridge wall connects. The secondary air passage is constructed in the bridge wall and extends out through the exterior wall. The secondary-air-entry ports are located on the mixing-chamber side of the bridge wall, just below the flame port. If additional secondary air is needed, supplementary ports may be installed on the exterior wall of the mixing chamber.

Auxiliary Heat

Successful performance of an incinerator is dependent to a large degree on the attention given by the operator. Often inclusion of burners in the design of a unit is made in anticipation of problems with the human element. The use of burners to provide auxiliary heat in multiple-chamber incinerators is normally not required for the burning of type 1 refuse. If a burner is included, it need be provided only in the mixing chamber. Auxiliary burners are not usually needed in the ignition chamber with this type of refuse, except possibly during initial lightoff or burndown periods.

The higher moisture content of type 2 refuse makes it much more difficult to burn. This often requires the addition of a burner in the ignition chamber as well as a secondary burner in the mixing chamber. The function of a burner is of course to provide auxiliary heat, when and if needed. The higher the moisture content of the refuse, the greater are the auxiliary heat requirements. Required burner capacities are given in Table 5–1 in terms of the moisture content of the refuse. The determination of the size of burners required should be based on the highest moisture content of refuse expected to be burned in the incinerator. Auxiliary burners should be fired with either natural gas or manufactured gas. Liquid-fuel-fired burners are generally unacceptable because of the many operating difficulties that are encountered with both burner and fuel.

For incinerators with rated capacities of up to 350 lb/hr the primary, or ignition-chamber burner, as well as the mixing-chamber burner, may be of the venturi type, equipped with a burner block and cage. The cage should provide openings for secondary air as well as access to flame-failure controls and electric-spark ignition. Primary as well as secondary burners for units of 350 lb/hr and more should be of the nozzle-mix blower type. Burners of this type provide maximum flame coverage for a given quantity of gas consumed. Burner adjustments should be made to give a "luminous" flame and to provide flame coverage over the entire cross-sectional area of the mixing chamber. Flame-failure controls should be provided on all burners.

Stack Draft

The normal method of producing a negative pressure within the ignition chamber is by the use of a natural-draft stack that utilizes the buoyancy of the hot flue gases. Draft produced in this manner is directly related to the height of the stack and to the difference in the reciprocals of the absolute temperature of the flue gas and ambient air. The theoretical draft requirements of a stack may be calculated from the following formula [3]:

$$D_t = 0.52PH \frac{1}{T_0} = \frac{1}{T_a},$$

where D_t = theoretical draft in inches water column,
$\quad P$ = barometric pressure in pounds per square inch,
$\quad H$ = height of stack above breeching in feet,
$\quad T_0$ = ambient temperature in degrees Rankine,
$\quad T_a$ = average stack temperature in degrees Rankine.

The velocity of the effluent in the stack and the cross-sectional area of the stack affect the usable or available draft. As the velocity within the stack increases or its cross-sectional area decreases, the losses due to friction increase proportionately. This reduces available draft. Draft losses can be calculated from the following formulas [2]:

Frictional losses (round stacks):

$$F_s = \frac{0.008H(V)^2}{(D)(T)},$$

where F_s = friction loss in inches water column,
$\quad H$ = height of stack above breeching in feet,
$\quad V$ = velocity in feet per second,
$\quad D$ = stack diameter in feet,
$\quad T$ = temperature in degrees Rankine.

Frictional losses (rectangular stacks):

$$F_s = \frac{0.002H(V)^2}{(m)(T)},$$

where m is the hydraulic radius in feet,

$$m = \frac{\text{area}}{\text{wetted parimeter*}},$$

and the other symbols are as defined above.

* The distance around a rectangular stack.

Expansion losses (usually negligible):

$$F_e = \frac{0.012V^2}{T},$$

where F_e is the head loss in inches water column.

The draft developed by the stack must be sufficient to overcome frictional losses and leave a net draft available of from 0.05- to 0.10-in. water column negative pressure for the inspiration of combustion air through the primary air ports. The range of total available stack draft required is from 0.12 in. water column for a 50-lb/hr incinerator to 0.3 in. water column for a 2000-lb/hr unit.

Many multistory buildings have incinerator stacks that are considerably higher than is necessary for providing draft. The reasons for this may be to discharge the incinerator effluent above the roof level, to prevent discharge of the effluent into the windows of adjacent buildings, or for some other equally valid reason. In this situation the draft produced by the stack is usually excessive and must be reduced to an acceptable level. There are a number of ways of overcoming this problem, one of which is to use a guillotine type of damper in the breeching from the incinerator to the stack. The disadvantage of this method is the need for constant adjustment, particularly during the lightoff period, to maintain the draft at a proper level.

The best solution to this problem is to install a barometric damper in the stack or stack breeching. Once the damper has been adjusted, it will automatically regulate and maintain the draft at a proper level. Draft in a stack is reduced by the barometric damper by virtue of the ambient air introduced, which produces several effects: It cools the stack gases, increases the gas velocities, and increases the frictional losses.

APPLICATION OF BASIC DESIGN PARAMETERS

Incinerator-design factors that are considered as most important include the ratios of combustion-air distribution, supplementary draft and temperature criteria, ignition-chamber length-to-width ratios, arch height, and grate loading, as well as the secondary-combustion-stage velocity and proportion factors. Some of these factors are functions of the desired hourly combustion rate and are expressed in empirical formulas, whereas others are values independent of incinerator size.

The values determined for these parameters are mean empirical values, accurate in the same degree as the experimental accuracy of the evaluation tests from which they were developed. The significance of exact figures is reduced also by the fluctuation of fuel composition and other miscellaneous operational conditions. For purposes of design permissible variations from the optimum mean values shown are plus or minus 10 percent, and velocities may deviate as much as 20 percent without serious consequences.

Table 5–1 lists the basic parameters, evaluation factors and equations, and gives the optimum values established for each. Although heat-release rates are not generally used as criteria for establishing volume relationships of multiple-chamber incinerators, for comparative purposes Figure 5–5 has been included for those accustomed to sizing combustion furnaces by these criteria. It may be seen that the values lie within the acceptable range of conventional combustion-furnace design. In the smaller multiple-chamber incinerators heat-release rates, including heat from auxiliary fuels, approximate 30,000 Btu/(ft³)(hr), and in the largest incinerator units heat-release rates are on the order of 15,000 Btu/(ft³)(hr).

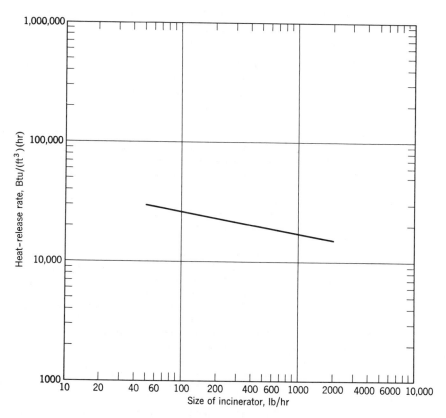

Figure 5–5 Heat-release rates for multiple-chamber incinerators.

The relationship shown in Table 5–1 may be applied generally to most commercial and industrial incinerators unless, as already indicated, the refuse is of an unusual nature or the conditions of operation are extremely variable, in which events special considerations must be made and the designs must be modified accordingly; for example, special problems are encountered when incinerator operation is intermittent or charging is not uniform; when the refuse contains such materials as plastics, rubber, or paint sludge; or when the refuse is extremely high in moisture.

Design Calculations

In order to use the design factors itemized in Table 5–1 calculations must be made that will yield incinerator data in usable form. The calculations fall into three general categories: (a) combustion calculations based on refuse composition, assumed air requirements, and estimated heat loss, (b) flow calculations based on the properties of the products of combustion and assumed gas temperatures, and (c) dimensional calculations based on simple mensuration and empirical sizing equations.

To facilitate the calculations of data needed to make comparisons with the recommended design parameters, certain assumptions are permissible. It is important, however, that these be reasonable estimates of actual expected conditions of operation. The simplifying assumptions on which calculations may be based are summarized as follows:

The burning rate and average refuse composition are taken as constant. The exception occurs when extremes in material quality and composition are encountered, in which case the most difficult burning conditions is assumed.

The average temperature of the combustion products is determined through normal heat-balance calculations. Heat losses due to radiation, refractory heat storage, and residue heat content may be assumed to average 20 to 30 percent of the gross heating value of the refuse during the first hour of operation. Furnace data generally available indicate that the losses approximate 10 to 15 percent of the gross heat after 4 to 5 hr of continuous operation.

The overall average gas temperature should be about 1000°F when calculations are based on 300 percent excess combustion air and the heat-loss assumptions previously given. The calculated temperature is not flame temperature and does not indicate the probable maximum temperatures attained in the flame port or mixing chamber. Should the temperature be lower, the need for auxiliary primary burners is indicated and should be sized as indicated in Table 5–1.

The temperatures used in checking gas-flow velocities are approximations of the actual temperature gradient in the incinerator since the products of combustion cool in passing from the flame port to the stack outlet.

Indraft velocities in the combustion-air ports (overfire, underfire, and secondary) are assumed to be equal, with a velocity pressure of 0.1 in. water column (equivalent to 1265 ft/min). Design the draft system so that available draft in the primary combustion chamber is about 0.1 in. water column and oversizing of adjustable air ports insures maintenance of proper air induction.

Air ports must be sized for admission of theoretical air plus 100 percent excess air. The remaining air enters the incinerator through the open charging door during batch operation and through expansion joints, cracks around doors, etc.

The combustion calculations needed to determine weights and velocities of the products of combustion and average temperatures may be derived from standard calculation procedures when the preceding assumptions are followed, using average gross heating values and theoretical air quantities. The sizing of air-inlet areas in the proportions designated is readily accomplished once the volumes of air and inlet velocities are established. The minimum areas required should be oversized in practice by the factor indicated in Table 5–1 in order to provide operational latitude.

Volume and temperature of the products of combustion are the only data that are needed to determine the cross-sectional flow areas of the respective ports and chambers. Calculations for draft characteristics follow standard stack-design procedures common to all combustion engineering. The stack velocity given for natural-draft systems is in line with good practice and minimizes flow losses in the stack.

The remainder of the essential calculations needed to design an incinerator are based on substitution in the parameter equations and measurements of the incinerator dimensions. Recommended grate loading, grate area, and average arch height may be calculated by equation or estimated from Figures 5–3 and 5–4. Proper length-to-width ratios may be determined and compared with proposed values.

Supplementary computations are usually required in determining necessary auxiliary-gas-burner sizes, stack draft control, and auxiliary-fuel-line piping. If moisture content of the refuse is less than 10 percent by weight, burners usually are not required. A moisture content of 10 to 20 percent normally indicates the necessity for installation of mixing-chamber burners, and a moisture content of more than 20 percent usually indicates that ignition-chamber burners also must be included.

These design criteria are reliable guidelines to the planning of combustible-refuse burners. The allowable deviations given in Table 5–1 should be interpreted with discretion to avoid consistently high or low deviation from the optimum values. Application of these factors to design evaluation must be tempered by judgment and by an appreciation of the practical limitations of construction and economy.

The following example shows the mathematical calculations necessary to design an incinerator:

PROBLEM.   Design a multiple-chamber incinerator to burn 100 lb/hr of paper with 15 percent moisture.

Solution.

1. Combustion of refuse:

   Dry combustibles (100 lb/hr)(0.85)   = 85 lb/hr.

   Moisture   (100 lb/hr)(0.15)   = 15 lb/hr.

2. Gross heat of combustion:

   From Appendix 5–I the gross heating value of
   dry paper is 7590 Btu/lb. Thus

   (85 lb/hr)(7590 Btu/lb)   = 645,200 Btu/hr.

3. Heat losses:

   From Appendix 5–I there are 0.56 lb of water
   formed from the combustion of 1 lb of dry paper.

   Radiation, etc. (assume 20 percent loss):

   (645,200 Btu/hr)(0.20)   = 129,040 Btu/hr.

   Evaporation of contained moisture:

   (15 lb/hr)(1060 Btu/lb)   = 15,900 Btu/hr.

   Evaporation of water from combustion:

   (0.56 lb/lb)(85 lb/hr)(1060 Btu/lb)   = 50,400 Btu/hr.

   Total = 195,340 Btu/hr.

4. Net heat:

   645,200 Btu/hr − 195,340 Btu/hr   = 449,860 Btu/hr.

5. Weight of combustion products with 300 percent excess air:

   From Appendix 5–I there are 21.7 lb of combustion products from the combustion of 1 lb
   of paper with 300 percent excess air.

   Paper (85 lb/hr)(21.7 lb/lb)   = 1844 lb/hr.

   Water (15 lb/hr)   =   15 lb/hr.

   Total = 1859 lb/hr.

6. Average gas temperature:

   The specific heat of the products of combustion is approximately 0.26 Btu/(lb)(°F).

$$T = \frac{Q}{C_p M},$$

where T = temperature difference in degrees Fahrenheit,

   Q = net heat in Btu,

   C_p = specific heat in Btu per pound per degree Fahrenheit,

   M = weight in pounds.

$$T = \frac{449,860 \text{ Btu/hr}}{[0.26 \text{ Btu/(lb)(°F)}](1859 \text{ lb/hr})} = 930°F,$$

$$T = T + 60°F,$$

$$T = 930 + 60 \qquad = 990°F.$$

7. Combustion-air requirements:
   (Basis: use 300 percent excess air; 200 percent excess air is admitted through open charging door and leakage around doors, ports, expansion joints, etc.)
   From Appendix 5–I, 68.05 ft^3 of air is theoretically necessary to burn 1 lb of dry paper.

   (85 lb/hr)(68.05 ft^3/lb)(2)   = 11,680 cfh,

   or = 192.3 cfm,

   or =   3.2 cfs.

8. Air-port-opening requirements at 0.1-in. water column: 1265 fpm is equivalent to a velocity pressure of 0.1 in.

$$\text{Total} = \frac{(192.3 \text{ cfm})(144 \text{ in.}^2/\text{ft}^2)}{1265 \text{ fpm}} \qquad = 22.0 \text{ in.}^2$$

   Air supply from Table 5–1: overfire, 70 percent; underfire, 10 percent; secondary, 20 percent.

   Overfire-air port (0.7)(22.0 in.^{2})   = 15.4 in.^{2}
   Underfire-air port (0.1)(22.0 in.^{2})   =   2.2 in.^{2}
   Secondary-air port (0.2)(22.0 in.^{2})   =   4.4 in.^{2}

9. Volume of products of combustion:
   From Appendix 5–I, 283.33 ft^3 of products of combustion are formed from the combustion of 1 lb of paper with 300 percent excess air.
   (Basis: 60°F and 300 percent excess air.)

   Paper (85 lb/hr)(283.33 ft^3/lb)   = 24,080 cfh.

$$\text{Water } (15 \text{ lb/hr}) \frac{379 \text{ ft}^3/(\text{lb})(\text{mole})}{18 \text{ lb/mole}} = \frac{316 \text{ cfh},}{24,396 \text{ cfh},}$$

   or =   6.8 cfs.

10. Volume of products of combustion through flame port:
    Total volume minus secondary air,
    6.8 cfs − (3.2 cfs)(0.20)   = 6.16 cfs.

11. Flame-port area:
    From Table 5–1 velocity is 55 fps. Thus

$$\frac{(6.16 \text{ cfs})(1560°R)}{(55 \text{ fps})(520°R)} \qquad = 0.34 \text{ ft.}^2$$

12. Mixing-chamber area:
    From Table 5–1 velocity is 25 fps. Thus

$$\frac{(6.8 \text{ cfs})(1460°R)}{(25 \text{ fps})(520°R)} \qquad = 0.76 \text{ ft.}^2$$

13. Curtain-wall-port area:
    From Table 5–1 velocity is 20 fps. Thus
    $$\frac{(6.8 \text{ cfs})(1410°\text{R})}{(20 \text{ fps})(520°\text{R})} = 0.92 \text{ ft.}^2$$

14. Combustion-chamber area:
    From Table 5–1 velocity is 6 to 10 fps (use 6 fps). Thus
    $$\frac{(6.8 \text{ cfs})(1360°\text{R})}{(6 \text{ fps})(520°\text{R})} = 2.96 \text{ ft.}^2$$

15. Stack area:
    From Table 5–1 velocity is 30 fps (use 25 fps). Thus
    $$\frac{(6.8 \text{ cfs})(1360°\text{R})}{(25 \text{ fps})(520°\text{R})} = 0.71 \text{ ft.}^2$$

16. Grate area:
    From Figure 5–4 the grate loading for average refuse is 18 lb/(ft²)(hr).
    $$\frac{(100 \text{ lb/hr})}{18 \text{ lb/(ft}^2)(\text{hr})} = 5.56 \text{ ft.}^2$$

17. Arch height:
    From Figure 5–3 the arch height   = 27 in.

18. Stack height [3]:
    Interpolated from Table 5–1, $D_t = 0.17$ in. water column.

$$D_t = 0.52PH\left(\frac{1}{T_0} - \frac{1}{T_a}\right),$$

where D_t = draft in inches water column,
      P = barometric pressure in pounds per square inch,
      H = height of stack above grates in feet,
      T_0 = ambient temperature in degrees Rankine,
      T_a = average stack temperature in degrees Rankine.

$$H = \frac{D_t}{(0.52)(P)[(1/T_0) - (1/T_a)]}$$
$$H = \frac{0.17}{(0.52)(14.7)(\frac{1}{520} - \frac{1}{1360})} = 18.75 \text{ ft.}$$

The following example illustrates the procedure for calculating the proper size of barometric damper required on the stack breeching, assuming that the incinerator described in the foregoing example was to be installed in the basement of a multistory building with an extremely tall stack.

PROBLEM.   Calculate the size of a barometric damper to be installed in the breeching between a basement 100-lb/hr multiple-chamber incinerator and the 92-ft stack to limit the draft in the combustion chamber of the multiple-chamber incinerator to 0.2 in. water column.

Given. The stack is 18 in. square and has a cross-sectional area of 2.25 ft². The stack extends 92 ft above the breeching. The breeching itself is a 12-in.-diameter insulated straight duct 10 ft long.

Solution.

1. Compute the theoretical draft in the breeching at average gas temperatures [3]:

$$D_t = 0.52PH \left(\frac{1}{T_0} - \frac{1}{T_a} \right),$$

where the symbols are as defined in the preceding problem.
For an average stack-gas temperature of 100°F

$$D_t = (0.52)(14.7)(92) \left(\frac{1}{520} - \frac{1}{560} \right) = 0.092 \text{ in. water column.}$$

Theoretical draft (calculated by the above formula) versus temperature is given in the following table:

Temperature (°F)	D_t (in. water column)	Temperature (°F)	D_t (in. water column)
100	0.092	400	0.53
200	0.29	500	0.62
300	0.43	600	0.69

2. Compute the weight of air that must enter through the barometric damper to cool the products of combustion from the multiple-chamber incinerator, neglecting heat losses. Assume a temperature of 300°F for first calculation. Although neglecting heat losses will cause the damper to be somewhat oversized, the draft can still be regulated with the weights on the damper. However, with the damper undersized, the draft cannot always be controlled.

$$(W_A)(C_{p2})(T_2 - T_A) = (W_{pc})(C_{p1})(T_1 - T_2),$$

where W_A = weight of air entering through barometric damper in pounds per second,

W_{pc} = weight of stack gases in pounds per second,

C_{p1} = average specific heat of products of combustion from multiple-chamber incinerator over temperature range of T_1 to T_2, in Btu/(lb)(°F),

C_{p2} = average specific heat of air over temperature range T_2 to T_A, in Btu/(lb)(°F),

T_2 = final temperature of stack gases, in degrees Fahrenheit,

T_1 = average temperature of gases from multiple-chamber incinerator, in degrees Fahrenheit,

T_A = temperature of air, in degrees Fahrenheit.

From the illustration given earlier the mass flow rate of the products of combustion through the multiple-chamber incinerator is 1859 lb/hr, the temperature is 990°F, and the specific heat is 0.26 Btu/(lb)(°F). The specific heat of ambient air is 0.24 Btu/(lb)(°F).

$(W_A$ lb/sec)[0.24 Btu/(lb)(°F)](300°F − 60°F)

$\qquad$ = (1859 lb/hr)[0.26 Btu/(lb)(°F)](990°F − 300°F)

$$W_A = 1.61 \text{ lb/sec.}$$

3. Volume of air entering through the barometric damper:

$$\text{Volume } (300°F) = \frac{(1.61 \text{ lb/sec})(379 \text{ ft}^3/\text{mole})(760°R)}{(29 \text{ lb/mole})(520°R)} = 30.8 \text{ cfs.}$$

4. Volume of products of combustion from the multiple-chamber incinerator at 300°F:

From the illustrative problem the volume of products of combustion is 6.8 cfs at 60°F.

$$\text{Volume } (300°F) = \frac{(6.8 \text{ cfs})(760°R)}{520°R} = 9.93 \text{ cfs.}$$

5. Total volume flowing through the breeching and stack at 300°F:

$$\text{Total volume } (300°F) = 30.8 + 9.93 \text{ cfs} = 40.73 \text{ cfs.}$$

6. Velocity through the breeching:

Area of breeching = 0.78 ft².

$$\text{Velocity } (300°F) = \frac{40.73 \text{ cfs}}{0.785 \text{ ft}^2} = 51.9 \text{ fps.}$$

7. Friction loss in the breeching [2]:

$$F_B = \frac{0.008(H)(V)^2}{(D)(T)},$$

where F_B = friction loss in inches water column,
H = length of breeching in feet,
V = velocity in feet per second,
D = duct diameter in feet,
T = temperature in degrees Rankine.

$$F_B = \frac{(0.008)(10)(51.9)^2}{(1)(760)} \qquad = 18.3 \text{ fps.}$$

8. Friction loss in the stack [2]:

$$F_S = \frac{0.002(H)(V)^2}{(m)(T)},$$

where F_S is loss in inches water column and m is the hydraulic radius in feet.
For rectangular cross section the hydraulic radius is

$$m = \frac{(2.25 \text{ ft}^2)(12 \text{ in./ft})}{(4)(18 \text{ in.})} \qquad = 0.375 \text{ ft.}$$

$$F_S = \frac{(0.002)(92)(18.3)^2}{(0.375)(760)} \qquad = 0.216 \text{ in.}$$

water
column.

9. Total friction losses in breeching and stack:
Total friction losses = 0.283 + 0.216 in. water column = 0.499 in.

water
column.

10. Frictional losses (calculated by the above method with additional cooling air through the damper) for assumed stack-gas temperatures of 400 and 500°F are given in the following table:

Temperature (°F)	Friction Loss (in. water column)
300	0.50
400	0.28
500	0.18

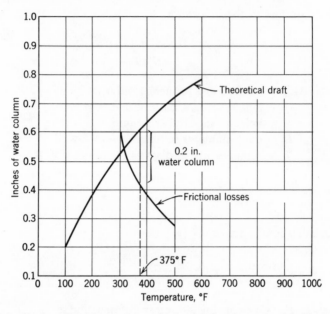

Figure 5–6 Draft at breeching of a multiple-chamber basement installation versus average stack-gas temperature.

11. Determine the stack-gas temperature:
    A stack-gas temperature of 375°F, representing a difference of 0.2 in. water column between the theoretical draft and the frictional losses, is obtained from a plot of the data derived herein, as shown in Figure 5–6.

12. Weight of air entering through the barometric damper at 375°F:

$$(W_A)(C_{p2})(T_2 - T_A) = (W_{pc})(C_{p1})(T_1 - T_2),$$

where W_A = weight of air entering through barometric damper in pounds per second,

W_{pc} = weight of stack gases in pounds per second,

C_{p1} = average specific heat of products of combustion from multiple-chamber incinerator over temperature range of T_1 to T_2 in Btu per pound per degree Fahrenheit,

C_{p2} = average specific heat of air over temperature range T_2 to T_A in Btu per pound per degree Fahrenheit,

T_2 = final temperature of stack gases in degrees Fahrenheit,

T_1 = average temperature of gases from multiple-chamber incinerator in degrees Fahrenheit,

T_A = temperature of air in degrees Fahrenheit.

$$(W_A \text{ lb/sec})[0.24 \text{ Btu/(lb)(°F)}](375°F - 60°F)$$
$$= (0.517 \text{ lb/sec})[(0.26 \text{ Btu/(lb)(°F)}](990°F - 375°F)$$

$$W_A = 1.10 \text{ lb/sec.}$$

13. Volume of air entering through the barometric damper:

$$\text{Volume } (60°F) = \frac{(1.10 \text{ lb/sec})(60 \text{ sec/min})(379 \text{ ft}^3 \text{ mole})}{29 \text{ lb/mole}}$$

$$= 864 \text{ cfm.}$$

14. Area of barometric damper:

The effective open area of a barometric damper is about 70 percent of its cross-sectional area. The area based on the calculated amount of air to be inspirated must therefore be increased accordingly. One velocity head at 0.2 in. water column and 60°F is 1766 fpm.

$$\text{Area} = \frac{(864 \text{ cfm})(144 \text{ in.}^2/\text{ft}^2)(1.3)}{1766 \text{ fpm}} = 101 \text{ in.}^2$$

15. Diameter (sized to the nearest inch) of barometric damper:

$$\text{Area} = \frac{(\pi)(\text{diameter})^2}{4}$$

$$\text{Diameter} = \left[\frac{(4)(101)}{3.14} \right] = 12 \text{ in.}$$

GAS SCRUBBERS*

Gas scrubbers employing water sprays are sometimes used in connection with incinerator operations in order to condition the effluent for one reason or another. Their purpose usually is to cool the effluent to a temperature low enough so that an induced-draft fan may be used to replace a stack or else to remove large fly-ash particles, which may be a source of nuisance to adjacent buildings or to other facilities. The effluent from a multiple-chamber incinerator may contain some fly ash with particle sizes in excess of 60 microns. When it is required that these

* See Chapter 3 for theoretical treatment of wet scrubbers.

particles be completely eliminated, a scrubber can be employed. Scrubbers can readily be designed to handle both the task of cooling and fly-ash removal. The cost of such installations is not prohibitive.

There are several requirements associated with scrubbers that must be considered whenever their use is contemplated. These include electric power to operate the induced-draft fan, fresh water to cool the effluent, and a means of disposing of contaminated water from the scrubber. In some areas provision must be made for a clarifier to remove fly ash and other collected solids before the water is discharged to a sewer. In addition to these basic requirements there are several other considerations inherent in the use of a scrubber. Most important is the increased maintenance due to corrosion resulting from condensed acids. It is for this reason that scrubber water is seldom recirculated. Even when the scrubber is lined with a dense refractory material, acid attack on the exterior steel shell is almost inevitable. In addition there is corrosion and erosion of both the fan impeller and fan housing. Also the prolonged contact of acidic water in the sump of the scrubber will result in its deterioration.

There are several basic considerations that are important in designing a gas scrubber. Perhaps the most important is that there be no carry-over of water in the effluent discharged from the fan. The deposit of acidic water droplets in the area adjacent to the scrubber could prove to be a real nuisance. In order to prevent this occurrence the scrubber design should provide sufficient residence time to completely vaporize the water entrained in the effluent. To accomplish this the water-gas mixture should be retained within the scrubber from 1 to $1\frac{1}{2}$ sec and velocities should not exceed 15 ft/sec.

In considering the overall dimensional factors for scrubbers certain arbitrary guidelines have been used; for example, from an aesthetic standpoint it is desirable to have a scrubber that is neither longer nor higher than the incinerator. Also, the width of the scrubber should be such that it can be located either adjacent to or at the rear of any retort type of incinerator. The best location for a scrubber serving an in-line type of incinerator is at the rear of the final combustion chamber; however, it is entirely permissible and feasible to locate the scrubber adjacent to this chamber.

The design parameters recommended for gas scrubbers are summarized in the following:

1. The water rate to the scrubber should be about 1 gpm for every 100 lb/hr of rated incinerator capacity. This gives a water-to-gas ratio of about 1 gpm per 400 scfm of effluent.

2. The exhaust fan should be designed to handle 700 cfm at 350°F for every 100 lb/hr incinerator capacity.

3. The fan should be designed to provide $\frac{1}{2}$-in. static pressure for a 50-lb/hr incinerator, uniformly increasing to $1\frac{1}{2}$ in. for a 2000-lb/hr incinerator. These static pressures should be developed with the fan operating at 350°F. The static pressure developed by fans operating at 350°F is approximately two-thirds of that developed when handling air at ambient temperatures. When selecting an induced-draft fan from manufacturer's catalog table values, one should be chosen that will develop static pressures 50 percent higher than those given above, or $\frac{3}{4}$ in. for a 50-lb/hr incinerator to $2\frac{1}{4}$ in. for a 2000-lb/hr incinerator.

4. The horsepower requirements of the fan should be based on the full capacity of the fan at ambient temperature, not at 350°F.

The internal sizing of chambers and ports in a scrubber can be determined from the data given in Figure 5–7. Figures 5–8 and 5–9 present dimensional standards for gas scrubbers capable of effectively handling the effluent from retort and in-line types of multiple-chamber incinerator.

In considering the overall design of scrubbers it has been determined that air dilution of the gases from the incinerator prior to entering the scrubber is unnecessary. Water should be introduced into the effluent as it enters the scrubber and flows concurrent down its first pass.

By immediately introducing water into the gas stream, more time is allowed for mixing and evaporation while the desired cooling is accomplished. The average velocity of the gas-water mixture in the first pass should range from 9 to 10 ft/sec. The velocity of the gases in the up-pass is determined by calculating the remaining time requirement so that the gases remain within the scrubber for a total time of approximately $1\frac{1}{4}$ sec. The curtain-wall port is sized in the velocity range from 18 to 20 ft/sec to prevent excessive pressure drop and to prevent water in the sump from being re-entrained in the effluent. The gases exit from the extreme top of the up-pass so that its full length can be used for the evaporation of any water remaining in the gas stream. This location also prevents water, which tends to travel up the back side of the scrubber, from becoming re-entrained in the gas stream. Another provision, which also reduces the possibility of re-entrainment of the larger diameter water droplets, is the inclusion of a 4-in. channel at the bottom of the curtain wall on the down-pass side. This channel prevents water that impinges on the wall from forming larger diameter droplets and dripping from the bottom of the curtain wall into the high-velocity gas stream below. The channel collects the larger droplets and carries the water across the width of the scrubber, down its side walls, and into the sump

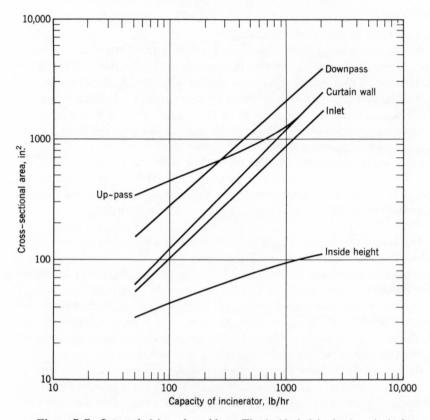

Figure 5–7 Internal sizing of scrubbers. The inside height is given in inches.

below. Additional structural support for the refractory in the dividing wall is also provided in this manner.

A level of water approximately 3 in. deep should be maintained across the base of the scrubber so that fly ash and other materials removed from the gas stream may be easily deposited and remain in this location. The water level is maintained by a large-diameter drainpipe that extends above the floor of the scrubber and into which the excess water overflows. The drainpipe should be removable at floor level so that fly ash and other solids can be washed down the sloping floor of the scrubber.

There are many types of automatic controls that may be used to regulate the temperature of the gases leaving the scrubber. Satisfactory controls that have proved to be both simple and economical include a hand-operated control valve and two solenoid valves, which operate automatically, installed in parallel between the water supply and the nozzles. Should the automatic-control system fail, the operator may open the hand valve and furnish sufficient water to the scrubber. The other valves are electrically connected so that one of the solenoid valves opens when the fan is placed in operation. The flow of water through this valve is adjusted to approximately 40 percent of the scrubber's needs. The other valve is controlled by a thermocouple located at the fan inlet. When the temperature at this point reaches 220°F the second solenoid valve opens, permitting the remainder of the water to be delivered to the nozzles. This simple arrangement furnishes adequate temperature control so that the fan's temperature never exceeds 350°F.

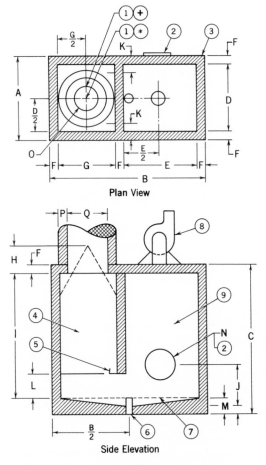

Plan View

Side Elevation

Figure 5–8 Design standards for retort incinerator scrubbers.

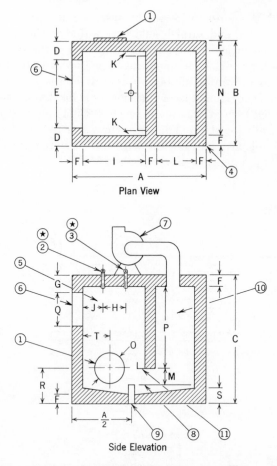

Figure 5–9 Design standards for in-line incinerator scrubbers.

A safety system also may be installed to prevent heat damage to the fan. The type most frequently used consists of a thermocouple located at the fan inlet and an additional solenoid valve that controls spray nozzles located in either the downpass or the up-pass of the scrubber. The nozzles should be capable of supplying at least the same quantity of water as the combined volume of the primary and secondary sprays. Should the temperature at the fan inlet exceed 550°F, a solenoid valve will open permitting the full volume of water to flow to the nozzles, cooling the effluent to an acceptable level. Unless these controls are tested frequently, when their need arises they will usually be inoperable. Consequently a safety system of this type is effective only if a regular maintenance and testing program is followed.

The induced-draft fan should be constructed of mild steel and be capable of withstanding temperatures up to 600°F. It is advisable that the fan be capable of at least two-speed operation; or better yet, that it have a variable-speed drive that may be adjusted from maximum capacity to one-third of maximum delivery volume. Controls of this type permit the operator to reduce the volume handled by the fan when the incinerator is operating at less than rated capacity. This, in turn, will increase the operating temperature within the incinerator and reduce the possibility of water carry-over from the scrubber. The fan controls should be readily accessible to the operator. This will facilitate adjustments in fan speed and increase the overall efficiency of the incinerator.

The exterior shell of the scrubber should be constructed of $\frac{3}{16}$-in. steel plate. Hangers should be mounted on the walls and on top of the scrubber on 9-in. centers to hold the lining firmly to the walls. Linings of 135-lb/ft^3 castable refractory should be 3 in. thick for incinerators with capacities of 750 lb/hr or less. Units with capacities in excess of 750 lb/hr should utilize 4-in. linings. The castable refractory floor should be sloped upward from the center of the scrubber at a 4° angle to facilitate the removal of collected fly ash and solids. The primary spray nozzles should be of the flat-spray type so that water droplets do not enter the connecting breeching and cause damage to the refractory in the final combustion chamber of the incinerator. The secondary nozzles should be of the full-cone type with a discharge angle of approximately 60°. Nozzles mounted within the inlet duct should be provided with an access opening for cleaning or replacement. Nozzles mounted at the top of the unit should be installed out of the hot gas stream and should be removable from the exterior of the scrubber. Nozzles should be constructed of brass or stainless steel.

Installation of mist eliminators is ordinarily not required; on occasion, however, water droplets may be discharged from the exhaust fan. Should this represent a serious problem, an eliminator section may be installed near the top of the up-pass. In general mist eliminators should be installed only after it has been found that the performance of an existing unit is unsatisfactory.

TYPES AND LIMITATIONS OF INCINERATOR DESIGN

Investigations of the performance characteristics of various multiple-chamber incinerator styles or types has resulted in the establishment of optimum design factors and operating limits for two basic types of multiple-chamber incinerators. The features that offer certain advantages in design, construction, and performance over a given operating range for each type are given below.

The essential features that distinguish the *retort* type of design are the following:

The arrangement of the chambers causes the combustion gases to flow through 90° turns in both horizontal and vertical directions.

The return flow of the gases permits the use of a common wall between the mixing and secondary combustion chamber.

Mixing chambers, flame ports, and curtain-wall ports have length-to-width ratios in the range from 1:1 to 2.4:1.

Bridge-wall thickness under the flame port is a function of dimensional requirements in the mixing and combustion chambers. This results in construction that is somewhat unwieldy in incinerators with capacities exceeding 500 lb/hr.

Distinguishing features of the *in-line* design are the following:

Flow of the combustion gases is straight through the incinerator with 90° turns in only the vertical direction.

The in-line arrangement of the component chambers gives a rectangular plan to the incinerator. This style is readily adaptable to installations that require separated spacing of the chambers for operating, maintenance, or other reasons.

All ports and chambers extend across the full width of the incinerator and are as wide as the ignition chamber. Length-to-width ratios of the flame-port, mixing-chamber, and curtain-wall-port flow cross sections range from 2:1 to 5:1.

Comparison of Types

A retort incinerator in its optimum size range offers the advantages of compactness and structural economy as the result of its cubic shape and minimal exterior wall length. It has been demonstrated that the retort incinerator performs more efficiently than its in-line counterpart in the capacity range of from 50 to about 750 lb/hr. The in-line incinerator is better suited to high-capacity operation than the retort but is less satisfactory for service in small sizes. The smaller in-line incinerators are somewhat less efficient with regard to secondary stage combustion than retort types. The in-line incinerator functions best at capacities that exceed 1000 lb/hr.

The efficiencies of both types in the capacity range between 750 and 1000 lb/hr are not too dissimilar. The choice of in-line or retort type in this capacity range is dictated by personal preference, space limitations, and, perhaps, the nature of the refuse and charging conditions.

The factors that tend to cause a difference in performance in the two incinerator types are (a) proportioning of the flame port and mixing chamber to maintain adequate gas velocities within the dimensional limitations imposed by the particular type involved, (b) maintenance of proper flame distribution over the flame port and across the mixing chamber, and (c) flame travel through the mixing chamber into the combustion chamber.

The additional turbulence and mixing of gases promoted by the directional changes in the retort-type incinerators allows the nearly square cross sections of the ports and chambers in small-size units to function adequately. In the retort sizes above 1000 lb/hr the reduced effective turbulence in the mixing chamber, caused by the increased size of the flow cross section, sometimes results in inadequate flame penetration, effluent distribution, and secondary air mixing.

As the capacity increases the in-line type exhibits structural and performance advantages. Some shortcomings of the small in-line type are eliminated as the size of the unit increases; for example, with an in-line incinerator of less than 750-lb/hr capacity the shortness of grate length in the ignition chamber tends to inhibit flame propagation across the width of the ignition chamber. This, coupled with thin flame distribution over the bridge wall (which is the wall separating the ignition and mixing chambers), may result in the passage of smoke from a smoldering fuel bed straight through the incinerator and out of the stack without adequate mixing and secondary combustion. In-line incinerators with capacities of 750 lb/hr or more have grates that are long enough, with respect to their overall width, to maintain burning across the full width of the ignition chamber. This results in satisfactory flame distribution in the flame port and mixing chamber. In the smaller in-line incinerators the relatively short grates add some construction problems.

Since the bridge wall usually is not provided with any structural support or backing and secondary air passages are built into it, it is very susceptible to structural failure. Careless stoking and grate cleaning in short-chambered in-line incinerators can demolish the bridge wall in short order.

No upper capacity limit has been given to the use of the in-line incinerator because little is known of operating efficiencies in capacities exceeding 8000 lb/hr. Incinerators with capacities less than 2000 lb/hr may be standardized for construction purposes to a large degree. Incinerators of larger capacity, however, are not readily standardized because problems of construction, materials, requirements for mechanized operation with stoking grates, induced-draft systems, and other factors make each such installation essentially one of custom design. The design factors advocated herein are equally applicable to the design of larger incinerators, up to and including municipal incinerators, as they are to the design of smaller units.

Standard Multiple-Chamber Design

Dimension standards for retort and in-line multiple-chamber incinerators are shown in Figures 5–10 and 5–11. These standards conform

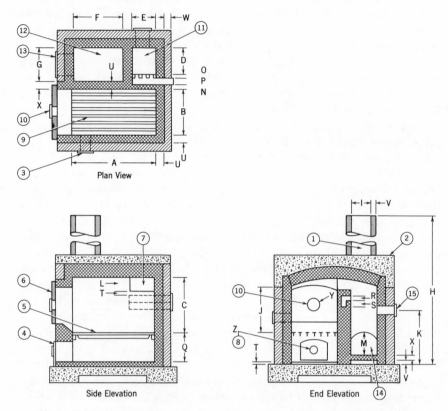

Figure 5-10 Design standards for retort type multiple-chamber incinerators.

to the recommended design criteria set forth in Table 5-1. They represent the optimum in design for these two basic types of incinerator. In the following are some provisional notes on the type of refuse considered in developing the standards and construction requirements for these units:

Refuse

Either type 1 or type 2.

Construction

1. Arch height and other vertical dimensions are averages based on construction with 60° sprung arches.

2. Sizes listed for circular ports are nominal diameters of round air-inlet spinners that have 35 to 50 percent net open-face area. Allowances must be made to provide equivalent areas in other air-port styles.

3. Refractory specifications are as follows:

Location	Brick	Castable
Walls, floor, and arches	High duty; PCE,[a] $32\frac{1}{2}$	120 lb/ft³; PCE, 17
Door linings	——	75 lb/ft³; PCE, 15
Stack linings	Insulating firebrick	75 lb/ft³; PCE, 15
Exterior insulation	2000°F	2000°F

[a] Pyrometric cone equivalent.

4. Grates shall be of **T** cross section, cast iron, and weigh not less than 40 lb/ft².

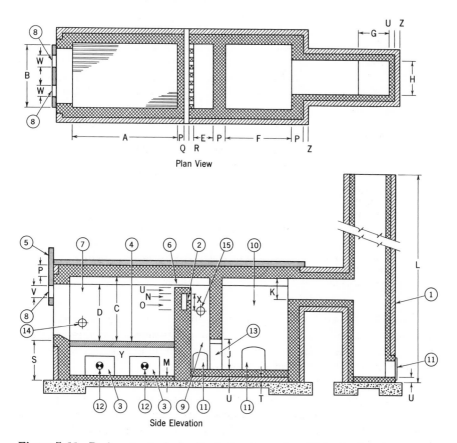

Plan View

Side Elevation

Figure 5-11 Design standards for the in-line type of multiple-chamber incinerator.

5. If exterior steel casings are used, they shall have a minimum thickness of 10 gage for all units with a capacity of less than 500 lb/hr. For units of 500 lb/hr and more, not less than $\frac{3}{16}$-in. steel plate shall be used.

6. All gas burners shall be of the nozzle-mix type with 100 percent safety controls, including automatic shutoff on main burners as well as on the pilots. Gas volumetric rates required for burners are specified in Table 5–2.

Table 5–2   Gas Volumetric Rates Required for Auxiliary Burners on Multiple-Chamber Incinerators

| | Size of Burner (ft³/hr) | | | |
| | Primary Gas Burner | | Secondary Gas Burner | |
Capacity of Incinerator (lb/hr)	Type 1 Refuse	Type 2 Refuse	Type 1 Refuse	Type 2 Refuse
50	125	250	150	175
100	200	500	225	275
150	250	600	300	375
250	300	700	500	600
500	500	1000	750	950
ᵃ750	500	1400	1000	1250
ᵇ750	700	1400	1000	1250
ᵃ1000	750	1500	1200	1000
ᵇ1000	800	1600	1200	1000
1500	1000	2000	1500	1900
2000	1500	3000	2000	2500

[a] Retort type.
[b] In-line type.

The dimensions listed in Figures 5–12 and 5–13 will vary in some cases from optimum values. This has been done in order to preclude the use of partial or "cut" firebrick in wall construction. All dimensions are nominal, since brick-construction tolerances vary.

Dimension allowances must be modified if the construction differs from that used as the basis for the standards (i.e., flat or suspended arches instead of sprung arches, square or rectangular stack cross sections instead of round). Average arch heights and recommended flow areas must be maintained.

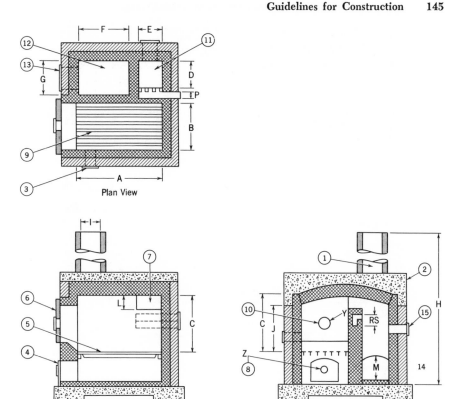

Figure 5–12 Dimensions of tested retort-type multiple-chamber incinerators.

External construction must be adequate structurally for the support of the refractory materials, with proper allowances for expansion and insulation. Foundation and stack-construction details should be in accord with good structural practices and in conformance with local building ordinances.

GUIDELINES FOR CONSTRUCTION

Mechanical design and construction of multiple-chamber incinerators are regulated in several ways. Ordinances and statutes establishing building codes that govern incinerator construction have been developed primarily on concepts of structural safety and fire prevention. Some air-pollution-control authorities have also specified certain material and

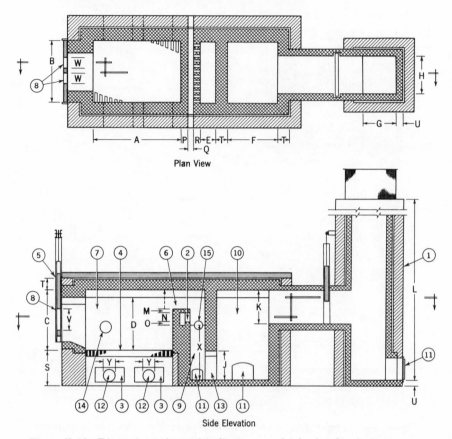

Figure 5–13 Dimensions of tested in-line-type multiple-chamber incinerators.

construction limitations that must be met. Incinerator manufacturers' associations have also recommended certain minimum standards to be followed. Many of the factors of construction are necessarily governed by the abrasion, erosion, spalling, and slagging encountered in the high-temperature ranges of incinerator operation.

The structural features and materials used in the construction of multiple-chamber incinerators are not readily discussed except in general terms. There are as many methods of erecting the walls of a multiple-chamber incinerator as there are materials from which to build them. The exterior walls of an incinerator may be constructed either of brick or steel plate. The refractory lining may be of firebrick, castable refractory, or plastic firebrick. The exterior walls must be protected from extreme-temperature conditions by providing either peripheral air space, air-cooling passages, or insulation. Stacks in small- to medium-size incinerators may be mounted directly on the incinerator, free standing, or integral within the building structure.

The most important element of multiple-chamber-incinerator construction, other than the basic design, is the proper installation and use of refractories. It is imperative that the manufacturer use suitable materials of construction and that these be installed by persons experienced in high-temperature-furnace fabrication and refractory installation. In choosing among available materials service conditions alone will dictate the type of lining needed for any furnace.

Refractory Walls, Linings, and Insulation

The minimum specifications for refractory materials used for lining the exterior walls of multiple-chamber incinerators are the following:

1. Types 1 and 2 refuse-burning service:
   (a) Firebrick; high heat duty—pyrometric cone equivalent not less than $32\frac{1}{2}$.
   (b) Castable refractory; not less than 120 lb/ft³—pyrometric cone equivalent not less than 17.
2. Wood, sawdust, and other high-temperature service:
   (a) Firebrick; superduty—pyrometric cone equivalent not less than 34.
   (b) Plastic—pyrometric cone equivalent not less than 34; not less than 130 lb/ft³.

Minimum refractory thickness for lining exterior walls (including arches) of incinerators burning all classes of refuse are the following:

1. Up to and including 350-lb/hr capacity: castable refractory or plastic—4 in.; firebrick—$4\frac{1}{2}$ in.
2. Above 350-lb/hr capacity: all refractories—9 in.

Stacks should be lined with refractory material with a minimum service temperature of 2000°F. In low-capacity units the minimum lining thickness should be $2\frac{1}{2}$ in.; in units larger than 350 lb/hr, $4\frac{1}{2}$ in.

Doors should be lined with refractory material with a minimum service temperature of 2800°F. Units smaller than 100 lb/hr should have door linings of 2-in. minimum thickness. In the size range of 100 to 350 lb/hr the linings should be increased to 3 in. In units with capacities from

350 to 1000 lb/hr the doors should be lined with 4 in. of refractory. In units of 1000 lb/hr and more linings should be 6 in.

The thickness of refractory lining and insulation in the floor of a multiple-chamber incinerator depends primarily on its physical location. Incinerators that are installed on their own concrete foundations outside of buildings should have $2\frac{1}{2}$ in. of firebrick lining backed by a minimum of $1\frac{1}{2}$ in. of 2000°F insulating material. Incinerators of the semiportable type should have sufficient air space provided beneath the incinerator so that no damage to the pad will result. When incinerators are installed within buildings it is extremely important that provision be made to prevent damage to floors and walls of the building. Such damage can be prevented by providing air passages beneath the incinerator and adjacent to the building walls to prevent excessive heat from actually reaching the structure. If an air space beneath the incinerator is impractical, then additional insulation should be provided.

For incinerators with capacities of up to 500 lb/hr $4\frac{1}{2}$ in. of firebrick and $2\frac{1}{2}$ in. of insulation should be provided on the floor of the mixing- and final-combustion chambers. For incinerators with capacities of 500 to 2000 lb/hr $4\frac{1}{2}$ in. of firebrick backed by 4 in. of insulation should be provided.

Units in high-temperature service should be provided with an insulating air space of 3 to 4 in. between the interior refractory and the exterior steel. This will reduce the temperature of the refractory and extend its life. Adequate openings above and below the incinerator should be furnished for air to enter and exit freely. In some cases forced circulation of the air in this space may be required. Some incinerator designs utilize forced preheated air as secondary combustion air.

The minimum thickness of interior refractory dividing walls generally follows that required for the exterior walls. The bridge wall, with its internal secondary-air-distribution channels, requires greater thickness. The minimum width of refractory material between the secondary air channel and the ignition or charging chamber should never be less than $2\frac{1}{2}$ in. for very small units, $4\frac{1}{2}$ in. for units up to 250 lb/hr, and 9 in. for larger units.

Expansion Joints

Sufficient expansion joints in the refractory construction are necessary to prevent buckling and destruction of the walls and arches. Each foot of wall made with firebrick-clay refractory will expand and contract between $\frac{1}{16}$ and $\frac{3}{32}$ in. when heated and cooled. Silica brick expands twice as much as fireclay brick. Provisions for vertical expansion should be sufficient between the arch and sidewalls to allow for vertical movement. Horizontal expansion has to be provided for between the various vertical walls. No hard and fast rules can be laid down for the provision of expansion joints. Their proper design requires complex calculation based on the experience of the contractor, along with engineering knowledge.

Steel Specifications

Where steel plate is used for the exterior casing of multiple-chamber incinerators, minimum thickness of the plate should be 10 gage for units of up to 500-lb/hr capacity. Units of 500 lb/hr and more require the use of $\frac{3}{16}$-in. plate steel. Exterior brick walls and steel casings of larger units must conform to minimum building-code structural requirements.

Stacks with steel exterior casings should be fabricated from a minimum of 10-gage steel. Stacks with an outside diameter of 3 ft should have steel casings made of $\frac{3}{16}$-in. plate. Casing thickness should be increased to $\frac{1}{4}$ in. for stacks with a diameter of 4 ft or more.

Grates and Hearths

Incinerators burning types 1 and 2 refuse may be equipped with cast-iron grates with T or channel cross section or 40-lb steel rails. As the size or capacity of an incinerator increases, the length of the ignition chamber also increases. This presents problems in the hand-fired incinerators with capacities exceeding 750 lb/hr because it becomes difficult to keep the rear section of the grates completely covered. The use of a solid hearth at the rear of the ignition chamber in such units is considered good practice. Hearths at this location prevent open areas from being formed in the refuse pile, which is usually thin at the rear of a long ignition chamber. The solid hearth prevents excessive underfire air from entering immediately in front of the bridge wall, which will result in both quenching of the hot gases and excessive carry-over of ash and unburned material into the mixing chamber. Installation of sloping grates (i.e., grates that slant down from the front to the rear of the ignition chamber) facilitates charging. The sloping grate results in an increased distance between the arch and the grate at the rear of the chamber, which assists in reducing fly-ash entrainment.

Combustion-Air Inlets

All combustion-air inlets should be of a type that provides positive control. Circular-spinner-type controls with rotating shutters are generally used for both underfire- and overfire-air openings in retort-type incinerators. These should be used only for the underfire-air openings in the in-line type of incinerator.

Rectangular ports with butterfly or hinged dampers should be used for all secondary air openings and for the overfire-air openings of in-line incinerators. All air-inlet controls should be of cast iron. Rectangular sliding dampers are not recommended because they have a tendency to warp and will inevitably become inoperative.

Foundations

Foundation requirements for all incinerators are determined by the weight of the incinerator and the soil conditions. The prefabricated, portable units are usually provided with sufficient air space between the floor and the foundation to prevent cracking. The on-site constructed units must be provided with either air spaces or an adequate layer of insulating material.

Supports

Prefabricated incinerators should have at least three heavy structural steel supports beneath their floors. Their purpose is to provide support for the three load-bearing walls of the unit and to permit it to be moved safely. Adequate support, without placing any of the load on the refractory walls, must be provided for any stack that is installed on top of an incinerator. The stack-support load must be carried by exterior load-bearing walls or by steel exterior wall supports.

INCINERATOR OPERATION

All of the necessary prerequisites for efficient combustion and minimization of contaminant discharge are provided in a properly designed multiple-chamber incinerator.

It is a characteristic of the well-designed multiple-chamber incinerator that emission control is built in. For all practical purposes the discharge of smoke or solid contaminants is almost entirely dependent on the actions of the operator. With refuse that has a low heating value or a high moisture content smoke is controlled by proper admission of combustion air and by proper utilization of secondary burners to maintain the efficiency of the secondary chamber. Proper functioning of this chamber depends on luminous-flame propagation, together with adequate temperatures necessary for gaseous-phase combustion. The need for the use of secondary burners is usually determined by observing the nature of flame travel from the ignition chamber and the extent of flame coverage at both the flame port and the curtain-wall port.

The most important single aspect of the operation of multiple-chamber incinerators is charging refuse into the ignition chamber. Proper charging is necessary to reduce emissions of fly ash, to maintain adequate flame coverage of the burning refuse pile and the flame port, and to prevent the fuel bed from thinning out at the rear of the ignition chamber, particularly in the large units.

Before any incinerator is placed into operation the grate and the ashpit beneath it should be cleaned. If the ashpit is allowed to become filled, overheating of the grates will occur, causing them to soften, bend, and even to fall from their mountings.

The secondary burner or burners should be ignited a few minutes before the incinerator is charged in order to preheat the secondary chambers. The charging and cleanout doors should remain closed, and the air ports should be open during the preheating period. If the flames from the secondary burners are driven upward on ignition and back through the flame port, instead of downward through the mixing chamber, there is a reverse draft condition, and the secondary burners should be shut off.

It is simple to overcome this problem by inserting a small piece of burning paper through the cleanout door into the combustion chamber. The cleanout door should then be closed and the secondary burners relit. The burning paper in the combustion chamber will provide enough heat at the right location to direct the movement of air up the stack and will initiate proper operation of the burners.

The initial charge of refuse should fill the ignition chamber to a depth of between one-half and three-quarters of the distance between the grates and the arch. Care should be exercised to keep the top of the pile below the flame-port opening. The initial charge should then be ignited at the top and rear of the pile at a point just below the flame-port opening, following which the charging door should be closed. The primary burner in the ignition chamber should be used only when the refuse is very moist. If the use of this burner is required, care should be taken to prevent blocking of the primary burner by the refuse pile. The overfire- and underfire-air ports are usually set at about the half-open position at lightoff; they should be opened gradually to their full-open position as the incinerator reaches stable operation at its rated burning capacity.

After approximately one-half of the initial charge of refuse has been burned, the remainder should be carefully stoked, if necessary. The burning refuse should then be pushed as far as possible to the rear of the grates. This operation must be performed carefully to prevent excessive emission of fly ash. Additional refuse may be subsequently charged to the incinerator. The new refuse should be charged onto the front end of the grates and never on top of the burning refuse pile. This method charging precludes smothering of the fire and maintains flame travel over the entire rear half of the ignition chamber, through the flame port, and well into the mixing chamber. Flames propagate evenly over the surface of the newly charged material, minimizing the possibility of smoke emission. This method of charging also minimizes the necessity of stoking or disturbing the burning pile, so that little or no fly ash is emitted. When all of the material to be burned has been charged into the incinerator the unit enters the burndown phase of its operation. When the last charge has been reduced to one-half or less of its original size all air-port openings to the incinerator should be set at the one-half-open position. The secondary burners are always left on until any smoke issuing from material remaining on the grates has ceased. At this time all burners may be shut off.

Although wood and sawdust burn at higher temperatures than other refuse, these same operating procedures in general are applicable. Auxiliary burners are not required for this type of refuse. Sawdust should be charged along with other materials at a rate approximating 10 to 15 percent by weight of the total amount of refuse charged. It is advisable to charge sawdust on top of other more bulky material so that it does not fall through the grates.

In operating incinerators burning only· paper or wood care should always be taken to insure that the burning pile at the rear of the grate does not become too thin. Should this happen, excessive underfire air admitted at this point will quench the hot gases entering the flame port, and incomplete combustion with excessive smoke will result. Use of secondary burners usually prevents incomplete combustion resulting from thin fuel beds.

Smoke emission around the charging door or ashpit door, or both, usually results from overcharging of refuse. The following steps in sequence should be taken to correct this condition:

1. If the primary burner is operating, shut it off.
2. Observe the burning pile and move any material that may be blocking the flame port.
3. Make sure that the cleanout door or doors in any of the secondary chambers of the incinerator are closed; the same applies to any air port on these doors.
4. Allow the fuel bed to burn down to normal operating depth and reduce the charging rate.

White smoke discharged from the incinerator stack is usually caused by the introduction of too much excess air. The following steps should be taken in sequence until the white smoke ceases:

1. Ignite the secondary chamber burner or check to see that it is still in operation.
2. Close the secondary air port or ports.
3. Close the underfire-air port.
4. Reduce the overfire-air-port opening.
5. If all secondary burner capacity is not being used, gradually increase the operating rate of the burner until full capacity is reached.
6. If all of these operations fail to stop the issuance of white smoke, examine the material to be charged. Very probably the white smoke is the result of finely divided mineral material present in the charge that is being carried out the stack. Such items as paper sacks that contain pigments or other metallic oxides or such minerals as calcium chloride, etc., cause white smoke.

Black smoke is usually caused by insufficient combustion air or a burning rate that is greatly in excess of the capacity of the incinerator. The following corrective steps should be taken until the issuance of black smoke ceases:

1. If the primary burner or burners is operating, shut it off.
2. Open the secondary air port or ports.
3. Open the overfire-air port.
4. Either ignite the secondary chamber burner or check to see that it is still in operation.
5. If the black smoke still continues, gradually open the charging door until it is approximately one-quarter open.
6. Should all of these steps fail to eliminate the black smoke, examine the material remaining to be charged. If highly volatile materials such as rubber, plastics, etc., are charged in too great a proportion to other less-volatile refuse, the resulting combustion rate may be too rapid for the incinerator to handle. Such materials may be charged in very small quantities along with other less-volatile refuse. With experimentation it is possible to determine the quantity that may be charged along with other materials. Generally, highly volatile materials must be charged at a rate of about 10 percent by weight of the total charge.

Figure 5–14 is provided to serve as a guide for operation of multiple-chamber incinerators. Posting of such guides on incinerators has proven of considerable value in promoting their proper operation.

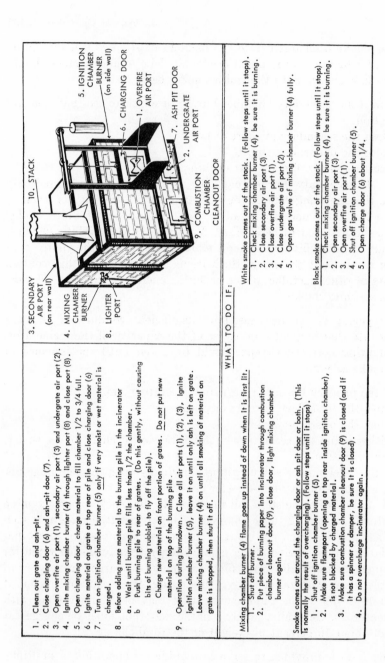

10. STACK
5. IGNITION CHAMBER BURNER (on side wall)
6. CHARGING DOOR
1. OVERFIRE AIR PORT
7. ASH PIT DOOR
2. UNDERGRATE AIR PORT
9. COMBUSTION CHAMBER CLEANOUT DOOR
3. SECONDARY AIR PORT (on rear wall)
4. MIXING CHAMBER BURNER
8. LIGHTER PORT

1. Clean out grate and ash-pit.
2. Close charging door (6) and ash-pit door (7).
3. Open overfire air port (1), secondary air port (3) and undergrate air port (2).
4. Ignite mixing chamber burner (4) through lighter port (8) and close port (8).
5. Open charging door, charge material to fill chamber 1/2 to 3/4 full.
6. Ignite material on grate at top rear of pile and close charging door (6)
7. Turn on ignition chamber burner (5) only if very moist or wet material is charged.
8. Before adding more material to the burning pile in the incinerator
   a. Wait until burning pile fills less than 1/2 the chamber.
   b. Push burning pile to rear of grates. (Do this gently, without causing bits of burning rubbish to fly off the pile).
   c. Charge new material on front portion of grates. Do not put new material on top of the burning pile.
9. Operation during burndown. Close all air ports (1), (2), (3), Ignite ignition chamber burner (5), leave it on until only ash is left on grate. Leave mixing chamber burner (4) on until all smoking of material on grate is stopped, then shut it off.

WHAT TO DO IF:

Mixing chamber burner (4) flame goes up instead of down when it is first lit.
1. Shut off burner.
2. Put piece of burning paper into incinerator through combustion chamber cleanout door (9), close door, light mixing chamber burner again.

Smoke comes out around the charging door or ash pit door or both. (This is normally the result of overcharging). (Follow steps until it stops)
1. Shut off ignition chamber burner (5).
2. Make sure flameport (opening at top rear inside ignition chamber), is not blocked by charged material.
3. Make sure combustion chamber cleanout door (9) is closed (and if it has a spinner or damper, be sure it is closed).
4. Do not overcharge incinerator again.

White smoke comes out of the stack. (Follow steps until it stops).
1. Check mixing chamber burner (4), be sure it is burning.
2. Close secondary air port (3).
3. Close overfire air port (1).
4. Close undergrate air port (2).
5. Open gas valve of mixing chamber burner (4) fully.

Black smoke comes out of the stack. (Follow steps until it stops).
1. Check mixing chamber burner (4), be sure it is burning.
2. Open secondary air port (3).
3. Open overfire air port (1).
4. Shut off ignition chamber burner (5).
5. Open charge door (6) about 1/4.

Figure 5-14 Operating instructions for multiple-chamber incinerator.

154

COST OF MULTIPLE-CHAMBER INCINERATORS

Many of the factors that affect the cost of most commodities are related to the cost of multiple-chamber incinerators. This applies specifically to commodities or services that are associated with the building trades. Usually the cost of the incinerator is not affected greatly by the type of materials used in the construction of the exterior walls, whether these be brick or steel, or in the inside linings of the incinerator, or whether castable refractories or firebrick is used. One exception to this is in the use of plastic firebrick, which may increase installation costs by as much as 40 percent. Usually plastic-firebrick linings are installed only for high-temperature service in multiple-chamber incinerators when the gross heating value of the refuse is expected to be more than 8500 Btu/lb.

Equipment and installation costs are increased considerably by the addition of such appurtenances as mechanical stoking grates, continuous-ash-removal systems, and automated feeding and charging mechanisms. Devices such as these represent as much as 40 percent of the cost of an incinerator.

Table 5-3 shows a comparison of the costs of multiple-chamber incinerators of various capacities. Also shown are gas-scrubber costs for various sizes of units. The figures do not include the cost of foundations, electrical wiring, and gas and water piping. Actual costs should be

Table 5-3 Approximate Costs of Multiple-Chamber Incinerators and Scrubbers[a]

Capacity of Incinerator (lb/hr)	Incinerator Cost	Gas-Scrubber Cost
50	$ 1,200	$1,100
100	1,700	1,500
150	2,000	1,800
250	2,700	2,200
500	6,000	3,100
750	9,500	3,800
1000	12,500	4,400
1500	20,000	5,600
2000	25,000	6,600

[a] 1966.

Table 5–4 Operating Conditions of Tested Multiple-Chamber Incinerators

							Test Number									
	1a	1b	2	3	4a	4b	5	6	7	8	9	10	11a	11b	12	13
Operating conditions	Normal[a]	Normal[b]	Normal	Normal	Normal	Burn-down	Normal	Normal	Normal	Normal	Normal	Normal	Light-off	Normal	Normal	Normal
Incinerator capacity (lb/hr)	50	50	150	250	350	350	750	1000	850	1000	1000	1000	1000	1000	2500	6000
Weight of refuse burned (lb)	30	26	150	203	350	350	713	770	650	940	1068	820	1055	700	3825	6500
Burning rate:																
Testing period (min)	38	34	60	66	60	70	60	55	60	60	60	60	60	60	101	60
Rate (lb/hr)	47	46	150	185	350	—	713	870	650	940	1068	820	1055	700	2300	6000
Percent of rated capacity	95	92	100	74	100	—	95	88	76	94	107	82	105½	70	92	100
Charging:																
Rate (lb/batch)	2–4	2–4	25	10–15	30	—	20	75	50–100	(d)	(d)	40	(d)	(d)	400	650
Time interval between charges (min)	2–4	2–4	10	5	5	—	2	6–7	3–5	(d)	(d)	3	(d)	(d)	10	10
Composition of refuse charged (percent)[e]:																
Paper	100 (15)	69 (15)	0	85 (20)	0	0	71 (15)	83 (15)	100 (15)	0	0	100 (15)	0	0	100 (10)	65 (10)
Garbage	0	31 (70)	0	15 (20)	0	0	17 (70)	17 (70)	0	0	0	0	0	0	0	0
Wood	0	0	100 (10)	0	100 (10)	100 (10)	12 (15)	0	0	100 (10)	100 (10)	0	100 (10)	100 (10)	0	35 (15)
Auxiliary fuel (scfh):																
Primary chamber burner	None	165	None	185	None	None	None	None	None	None	None	None	None	None	None	None
Mixing chamber burner	None	165	None	800	None	None	1125	2850	822	None	None	1390	None	None	(f)	None
Combustion air (percent of total supply):																
Primary air—overfire	85	45	80	40	55	(c)	79	50	80	20	55	54	50	50	60	70
Primary air—underfire	15	10	10	10	3	(c)	7	20	10	4	15	7	10	10	3	10
Secondary air—mixing chamber	0	45	10	50	42	(c)	14	30	10	76	30	39	40	40	37	20
Stack:																
Flow rate (scfm at stack conditions)	174	193	420	480	557	637	1970	2190	6300	3200	3300	2700	2200	2130	13400	27500
Moisture (percent)	8.3	13.2	10.0	14.9	10.4	3.0	10.8	12.0	4.4	7.0	8.5	7.8	9.9	6.4	5.7	11.9
Orsat analysis (percent):																
Carbon dioxide	4.8	6.4	9.6	9.3	8.4	3.2	6.0	7.4	2.4	5.8	7.0	5.6	10.6	7.8	2.2	6.3
Oxygen	13.8	6.3	10.4	4.1	11.2	17.2	12.6	9.9	18.0	14.7	13.4	13.9	8.4	11.4	18.3	9.4
Carbon monoxide	0.0	0.0	0.0	0.0	0.0	0.0	0.0	0.0	0.0	0.0	0.0	0.0	0.0	0.0	0.0	0.0
Nitrogen	81.4	87.3	80.0	86.6	80.4	79.6	81.4	82.7	79.6	79.5	79.6	80.0	81.0	80.8	79.5	84.3
Incinerator temperatures (°F):																
Ignition chamber	(c)	(c)	(c)	(c)	1996	1523	(c)	(c)	(c)	(c)	2403	(c)	(c)	(c)	(c)	(c)
Mixing chamber	(c)	1623	(c)	1996	(c)	(c)	(c)	(c)	(c)	1646	(c)	(c)	(c)	(c)	(c)	(c)
Combustion chamber	(c)	(c)	1309	1358	1358	969	1560	(c)	(c)	1585	1844	(c)	(c)	(c)	(c)	1400
Stack	1160	1475	1020	1600	1140	750	910	872	872	1000	1160	1080	2200	2130	(c)	(c)

[a] Without burners.
[b] With burners.
[c] Not available.
[d] Mechanized feed.
[e] Numbers in parentheses indicate moisture content of constituent.
[f] Oil at 2.5 gph.

Table 5-5 Emissions from Multiple-Chamber Incinerators

	Test Number															
	1a	1b	2	3	4a	4b	5	6	7	8	9	10	11a	11b	12	13
Maximum opacity of stack gases (percent)	10	0	5	80	0	0	45	10	20	0	0	0	100	0	15	0
Smoking time (min)	1	0	3	1.5	0	0	1.0	2.5	1	0	0	0	0.5	0	9	0
Range of opacity less than maximum (percent)																
Smoking time (min)	0	0	0	50	0	0	15-40	0	10-15	0	0	0	15	0	0	0
Total smoking time (min)	1	0	3	3	0	0	2.0	2.5	18	0	0	0	4.5	0	9	0
Particulates[a] (gr/scf of stack gas):																
At stack conditions	0.0987	0.058	0.0418	0.0852	0.0238	0.0130	0.075	0.083	0.047	0.052	0.122	0.060	0.129	0.119	0.0197	0.0920
At 12% carbon dioxide[b]	0.270	0.300	0.058	0.254	0.0381	0.0503	0.205	0.248	0.274	0.116	0.230	0.140	0.160	0.200	0.113	0.200
Particulates[c] (gr/scf of dry gas):	0.0677	0.0394	0.0212	0.0404	0.0234	0.0121	0.055	0.0453	0.0313	0.025	0.056	0.050	0.056	0.040	0.0093	0.0662
At 12% carbon dioxide	0.185	0.182	0.0265	0.125	0.0334	0.0455	0.13	0.119	0.131	0.053	0.096	0.120	0.062	0.062	0.057	0.126
Gaseous emissions (gr/scf):																
Carbon monoxide	(d)	(d)	0.0	(d)	0.0	0.005	(d)	(d)	(d)	0.0	(d)	(d)	0.02	0.0025	(d)	(d)
Nitrogen oxides[e]	(d)	(d)	0.0417	(d)	0.0321	0.0069	0.000017	(d)	0.0136	0.0278	0.056	0.020	0.107	0.055	(d)	(d)
Aldehydes[e]	(d)	(d)	0.003	(d)	0.002	0.0008	0.0000003	(d)	0.0015	0.002	0.004	0.001	0.005	0.004	(d)	(d)
Organic acids[e]	(d)	(d)	0.0166	(d)	0.021	0.023	(d)	(d)	0.034	0.009	0.016	0.0005	0.071	0.02	(d)	(d)
Sulfur dioxide[e]	(d)	(d)	(d)	(d)	(d)	(d)	(d)	(d)	0.028	(d)	(d)	(d)	0.005	0.00	(d)	(d)
Sulfur trioxide[e]	(d)	(d)	(d)	(d)	(d)	(d)	(d)	(d)	(d)	(d)	(d)	(d)	0.0054	0.0050	(d)	(d)

a Total.
b Carbon dioxide from refuse only.
c Dry collection only.
d Not available.
e Stack conditions.

157

within 10 percent of the basic cost of the same size incinerator anywhere within the United States for units up to and including 1000-lb/hr capacity.

TESTING OF MULTIPLE-CHAMBER INCINERATORS

Operating-condition and stack-emission data obtained from performance tests on typical multiple-chamber incinerators ranging in capacity from 50 to 6000 lb/hr are shown in Tables 5–4 and 5–5. For the most part the tests are representative of normal operation at equilibrium-temperature conditions. One lightoff and one burndown test are included for comparative purposes. Ordinarily the concentrations of contaminants discharge are lower during these two phases of operation.

Operating conditions and other relevant data on the incinerators tested are given in Table 5–4. The stack-observation and emission data obtained from the tests are shown in Table 5–5. The dimensions, configuration, and other design details on the incinerators tested are given in Figures 5–12 and 5–13. Although some of the dimensions may deviate slightly from the recommended design criteria, in each case the differences are the result of some specific requirement for the unit and usually are within permissible limits. Most of these data are self-explanatory; however, to facilitate their use supplementary information is given below.

Auxiliary Fuel

The auxiliary fuel used was natural gas with a gross heating value of 1100 Btu/ft^3 and a specific gravity of 0.65.

Combustion Air

The distribution of combustion air was determined by measuring the static pressure at the various air-entry ports and calculating the volume of air admitted at these locations. The percentage of overfire air shown includes only the air admitted through the overfire-air ports and does not include air admitted through the charging door when open for charging. For this reason the percentage of overfire air will be somewhat lower than that actually provided during the test. The figures giving the distribution of combustion air have been rounded off to the nearest 5 percent.

Emissions

The emissions of nongaseous contaminants are reported in Table 5–5 in four different ways. The first gives the concentrations of solid and condensible liquid materials emitted in terms of grains per standard cubic foot (scf) of gas at stack conditions and the second converts these values to a basis of 12 percent carbon dioxide. The carbon dioxide conversion factor includes only carbon dioxide formed from the burning of refuse and does not include carbon dioxide formed from the burning of auxiliary fuel. The third group of figures gives the grains of solid particulate matter per standard cubic foot of dry stack gas. This figure represents only the material collected by the dry portion of the collection train. Generally, this train consists of a small-diameter glass cyclone followed by an alundum thimble. This figure is then calculated to grains per standard cubic foot of dry gas corrected to 12 percent carbon dioxide to give a fourth set of values. Again, the carbon dioxide correction factor includes only that formed from the burning of refuse.

Carbon Dioxide Formed from Auxiliary Fuel

On those units in which gas burners were installed and operated during the test a complete velocity traverse was conducted with the burners in operation and with no refuse being consumed. This procedure is used to determine the quantity of carbon dioxide produced from the auxiliary fuel. Where fuel meters were available theoretical calculations were made from the fuel-consumption measurements and used as a cross-check with values determined from the stack analysis.

Flue-Factor and Sampling-Nozzle Determination

Prior to an actual test a preliminary velocity traverse was run with the incinerator operating at equilibrium conditions, burning refuse similar to that anticipated during the testing period. From these data a flue factor was established and the required sampling-nozzle diameter was selected.

Sampling and Laboratory Procedures

A portion of the stack gases was drawn through the sampling apparatus isokinetically during the test period. The sampling apparatus used in the majority of tests consisted of a glass cyclone and an alundum thimble, followed by three series-connected impingers in an ice bath.

After the test the cyclone and sample tubes were brushed into a tared beaker, fired at 105°F, cooled in a desiccator, and weighed. The tared alundum thimble was dried for 30 min at 105°F, and the weight of material collected was determined by difference. The solution in the impingers and washings was evaporated to dryness in a tared beaker at 105°F, cooled in a desiccator, and weighed. The sum of the first two weights was used as the total weight of combustion contaminants collected by the dry-sampling-train method. The sum of the three weights was used as the total weight of combustion contaminants (solid and condensible liquids). A liquid-displacement method employing an acidified saturated sodium sulfate solution was used to collect carbon dioxide samples. Integrated gas samples were obtained by withdrawing effluent from the stacks during the entire testing period. Orsat analyses were performed on these gas samples with standard Orsat equipment. Nitrogen was determined by difference.

REFERENCES

[1] George, R. E., "Effects of Design Factors on Stack Emissions from Multiple-Chamber Incinerators," *APCA Proceedings*, West Coast Section, March 25–26, pp. 92–101, 1957, Los Angeles.

[2] Griswold, J., *Fuels, Combustion and Furnaces*, McGraw-Hill, New York, 1946, p. 374.

[3] Kent, R. T., *Mechanical Engineer's Handbook*, 11th ed., Wiley, New York, 1963, pp. 6–104.

[4] Williamson, J. E., R. J. MacKnight, and R. L. Chass, *Multiple-Chamber Incinerator Design Standards*, Los Angeles County Air-Pollution-Control District, October 1960.

APPENDIX 5–I

Chemical Properties and Combustion Data for Paper, Wood, and Garbage

	Material			
	Sulfite Paper[a]	Average Wood[b]	Douglas Fir[c]	Garbage[d]
Analysis				
Constituent:				
Carbon	44.34	49.56	52.30	52.78
Hydrogen	6.27	6.11	6.30	6.27
Nitrogen		0.07	0.10	
Oxygen	48.39	43.83	40.50	39.95
Ash	1.00	0.42	0.80	1.00
Gross heating value[e] (Btu/lb)	7590	8517	9050	8820

Combustion Data

Constituent[f]:	Cubic Feet	Pounds	Cubic Feet	Pounds	Cubic Feet	Pounds	Cubic Feet	Pounds
Theoretical air	67.58	5.165	77.30	5.909	84.16	6.433	85.12	6.507
40 % saturation at 60°F	68.05	5.188	77.84	5.935	84.75	6.461	85.72	6.536
Flue gas with theoretical air:								
Carbon dioxide	13.993	1.625	15.641	1.816	16.51	1.917	16.668	1.935
Nitrogen	53.401	3.947	61.104	4.517	66.53	4.918	67.234	4.976
Water formed	11.787	0.560	11.487	0.546	11.84	0.563	11.880	0.564
Water (air)	0.471	0.023	0.539	0.026	0.587	0.028	0.593	0.029
Total	79.652	6.155	88.771	6.905	95.467	7.426	96.375	7.495
Flue gas with percent excess air as indicated:								
0	79.65	6.16	88.77	6.91	95.47	7.43	96.38	7.50
50.0	113.44	8.74	127.42	9.86	137.55	10.64	139.24	10.77
100.0	147.23	11.32	166.07	12.81	179.63	13.86	182.00	14.04
150.0	181.26	13.91	204.99	15.78	222.01	17.09	224.86	17.21
200.0	215.28	16.51	243.91	18.75	264.38	20.12	267.72	20.58
300.0	283.33	21.70	321.75	24.68	349.13	26.58	353.44	27.12

[a] Sulfite-paper constituents:
Cellulose ($C_6H_{10}O_5$) 84 %
Hemicellulose ($C_5H_{10}O_5$) 8 %
Lignin ($C_6H_{10}O_5$) 6 %
Resin ($C_6H_{10}O_5$) 2 %
Ash ($C_{20}H_{30}O_2$) 1 %
[b] Kent, R. T., *Mechanical Engineer's Handbook*, 11th ed., Wiley, New York, 1936, pp. 6–104.
[c] Kent, R. T., *Mechanical Engineer's Handbook*, 12th ed., Wiley, New York, 1950, pp. 2–40.
[d] Estimated.
[e] Dry basis.
[f] Based on 1 lb.

APPENDIX 5–II

Definitions

Breeching The connection between the incinerator and the stack.

Bridge wall A partition wall between the ignition and mixing chambers over which the products of combustion pass.

Burners Primary: A burner installed in the ignition chamber to dehydrate and ignite the material to be burned. Secondary: A burner installed in the mixing chamber to furnish supplementary heat and to maintain a minimum operating temperature.

Burning rate The amount of waste consumed, expressed as pounds per square foot of burning area per hour.

Capacity The amount of waste consumed in pounds per hour.

Combustion chamber Primary: Chamber where ignition and burning of the waste occur. Secondary: Chamber where the remaining solids, vapors, and gases from the ignition chamber are burned and settling of fly ash takes place.

Curtain wall A refractory partition that deflects the gases in a downward direction.

Damper A manual or automatic device used to regulate the rate of flow of gases through the incinerator. Barometric: A pivoted, balanced plate normally installed in the breeching and actuated by the draft. Guillotine: An adjustable plate normally installed vertically in the breeching, counterbalanced for easier operation. Butterfly: An adjustable, pivoted plate. Sliding: An adjustable plate normally installed horizontally or vertically in the breeching.

Draft The pressure difference between the incinerator (or any component part) and the atmosphere that causes the products of combustion to flow through the incinerator. Natural: The negative pressure created by the difference in density between the hot flue gases and the atmosphere. Induced: The negative pressure created by the action of a fan, blower, or ejector located between the incinerator and the stack.

Fly ash All solids including ash, charred paper, cinders, dust, soot, or other partially incinerated matter carried in the products of combustion.

Grate A surface with suitable openings to support the fuel bed and permit passage of air through the fuel. It is located in the ignition chamber and is designed to permit removal of the unburned residue. It may be horizontal or inclined, stationary or movable, and operated manually or automatically.

Hearth Cold drying: A surface on which wet waste material is placed to dry prior to burning. Hot drying: A surface upon which wet material is placed to dry prior to burning and having hot combustion gases pass successively under the hearth.

Heat-release rate The amount of heat liberated in the incinerator, usually expressed as Btu per cubic foot per hour.

Ignition chamber The chamber into which the refuse is charged and burned. The first of three chambers in a multiple-chamber incinerator. Also referred to as the *primary combustion chamber*.

Incinerator Equipment in which solid, semisolid, liquid, or gaseous combustible wastes are ignited and burned, the solid residues of which contain little or no combustible material.

Incinerator, multiple-chamber An incinerator consisting of three refractory-lined chambers interconnected by gas passages, or ducts, and designed in such a manner as to provide for maximum combustion of the material to be burned. Depending on the arrangement of the chambers, multiple-chamber incinerators are designated as in-line or retort types.

Mixing chamber   The chamber following the ignition chamber through which the gases flow downward. Secondary air is admitted here and gaseous-phase combustion continues.

Overfire air   Air introduced into the ignition chamber above the fuel bed.

Scrubber   Equipment for removing fly ash and other objectionable materials from the products of combustion by means of sprays, wet baffles, etc. Also reduces excessive temperatures of effluent and substitutes a fan for a tall stack.

Secondary air   Air introduced into the mixing chamber to aid in gaseous-phase combustion.

Settling chamber   Chamber designed to reduce the velocity of the gases in order to permit the settling out of fly ash.

Stack, or chimney   A vertical passage of refractory that conducts the products of combustion to the atmosphere.

Underfire air   Air introduced into the ignition chamber through the fuel bed.

6

CENTRAL INCINERATION
OF COMMUNITY WASTES

Harold G. Meissner

Engineering problems involved in designing facilities for incinerating
municipal refuse are similar to those encountered in designing boiler
plants or other fuel-burning equipment. When refuse is considered as a
fuel many of the misunderstandings and unsatisfactory performance con-
cerning municipal incineration are overcome. Incineration has been
defined as the disposal of waste or refuse by combustion for the purpose
of obtaining an inoffensive residue and effluent with a reduction of bulk.

The first step in selecting an incinerator for a community should be to
determine the average amount of refuse collected, its general characteristics, as well as its variations from day to day and by season. Such
information should come from existing records where possible; otherwise,
from a checkup made preferably during both warm and cold weather.
An average amount for many localities is 4.5 lb of refuse per day per
person, but this may vary, depending on the nature and location of the
pickup area. Factors such as the extent of industrialization and the type
of dwellings, private or multiple, are important [7].

Consultant and Professional Engineer. Formerly Assistant Director of Engineering,
Department of Air Pollution Control, New York City.

Table 6–1   Composition and Analysis of a Composite Municipal Refuse as Charged to Incinerator[a]

Constituent	Percent	Constituent	Percent
Corrugated-paper boxes	23.38	Ripe tree leaves	2.29
Newspaper	9.40	Flower garden plants	1.53
Magazine paper	6.80	Lawn grass, green	1.53
Brown paper	5.57	Evergreens	1.53
Mail	2.75	Plastics	0.76
Paper food cartons	2.06	Rags	0.76
Tissue paper	1.98	Leather goods	0.38
Plastic-coated paper	0.76	Rubber composition	0.38
Wax cartons	0.76	Paints and oils	0.76
Vegetable food wastes	2.29	Vacuum-cleaner catch	0.76
Citrus rinds and seeds	1.53	Dirt	1.53
Meat scraps, cooked	2.29	Metals	6.85
Fried fats	2.29	Glass, ceramics, ash	7.73
Wood	2.29	Extraneous moisture	9.05
			100.00

Proximate Analysis		Ultimate Analysis	
Constituent	Percent	Constituent	Percent
Moisture	20.00	Moisture	20.00
Volatile matter	52.70	Carbon	29.83
Fixed carbon	7.30	Hydrogen	3.99
Ash and metal	20.00	Oxygen	25.69
	100.00	Nitrogen	0.37
		Sulfur	0.12
		Ash and metal	20.00
			100.00

Heating value (Btu/lb): $5260 +$ [b]$182 = 5442$

Theoretical air required for combustion:

Carbon: 0.2983×11.53 = 3.442

Net hydrogen: $(0.0399 - 0.2569/8) \times 34.34$ = 0.264

50% of metals: 0.0686×1.05 = 0.072

Total pounds of dry air per pound of refuse or 50 ft³ of 65°F air   3.778

[a] Data from [4].

[b] From 50% oxidation of metals.

REFUSE

Characteristics

Municipal refuse is changing rapidly both in character and quantity. In general it is becoming bulkier and more combustible, because it is affected by rising standards of living and the trend of modern food processing toward packaging in combustible containers. The amount of dry combustible material such as paper containers, and plastic bags and wrappings, has increased greatly, whereas the amount of garbage and similar high-moisture ingredients has decreased. Table 6–1 gives analyses of a composite of present-day refuse as delivered; Figure 6–1 shows what it looks like on the dump floor. The preponderance of crates, cartons, containers, and similar bulky and highly combustible material is apparent. This radical difference in refuse analyses from those of a generation or less ago must be considered in planning pickup, handling, storage, and furnace facilities. Furthermore, as may be seen from Figures 6–2 and 6–3, daily and monthly variations in refuse collection affect incinerator-design parameters.

Objectives in Disposal

In recent years the heat value of a given weight of refuse has increased by at least 50 percent; the quantities of combustion air and gas, and the

Figure 6–1   Refuse on the dump floor.

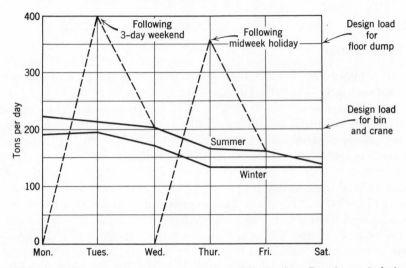

Figure 6–2 Daily variations in collection of combined refuse. Based on tabulations in *Refuse Collection Method Practice,* American Public Works Association, with modifications by the author.

furnace volume required have correspondingly increased. To incinerate means to burn to ashes. An efficient incinerator should accomplish this effectively and economically, as should any fuel-burning furnace. The incinerator should be designed to take practically any waste material that may be delivered to it, even though much of the material may be incombustible in whole or in part. Modern techniques in steel plants and paper mills have made salvage of metal and paper from refuse generally

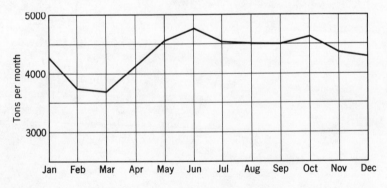

Figure 6–3 Monthly variations in the collection of combined refuse (Bethesda, Maryland, 1956).

unprofitable. Noncombustibles, such as bottles and cans, frequently present in refuse in large numbers, will form a major part of the residue, in spite of some oxidation of the metal containers in the hot furnace. Even with the tin cans and bottles left in, the reduction in bulk between refuse and residue will range from 10:1 to 20:1. The trend toward combustible containers will probably keep this ratio increasing.

The well-designed incinerator incorporates adequate provisions for controlling smoke, fume, and fly ash. As communities come to fully appreciate the harmful effects of polluted air and to realize that it is avoidable, they will pass stricter clean-air laws and regulations.

INCINERATOR PLANNING

Plant Location

The modern incinerator plant can have low outlines, relatively short stacks, and an attractive exterior. It should be as close as possible to the load center, preferably in an industrial section or in an area where the flow of trucks to and from the plant will not be too objectionable. Locating it on a hillside can be advantageous—achieving the several elevations required and space for storage bins with a minimum of ramp construction and excavation. Figure 6–4 shows a typical hillside installation and the elevations involved. Locating the incinerator close to a sewage-disposal plant makes it convenient to use the hot gas from the furnace to flash dry and sterilize the sewage effluent and burn the solids to inert ash. This effluent can also be used for spray water to cool and clean the combustion products when no heat-recovery equipment such as boilers or evaporators is installed.

Combined Facilities for Refuse and Sewage

The combination of refuse incineration and sewage disposal should be carefully evaluated; it may effect plant economies in both first cost and operating expenses, thus enabling many communities to provide modern disposal facilities where they would otherwise be impossible. Handling the combined disposal requirements for a number of nearby communities by the formation of sanitation districts or similar organizations offers many attractive advantages. The combustible gas that is generated in sewage treatment may be used to generate steam, and the hot gas from the incinerator may be used not only to flash dry and sterilize the sewage effluent but also to burn the solids to inert ash or dry and prepare them for use as fertilizer or soil-conditioner products. In many cases the effluent from the sewage treatment may be used instead of fresh water in the gas

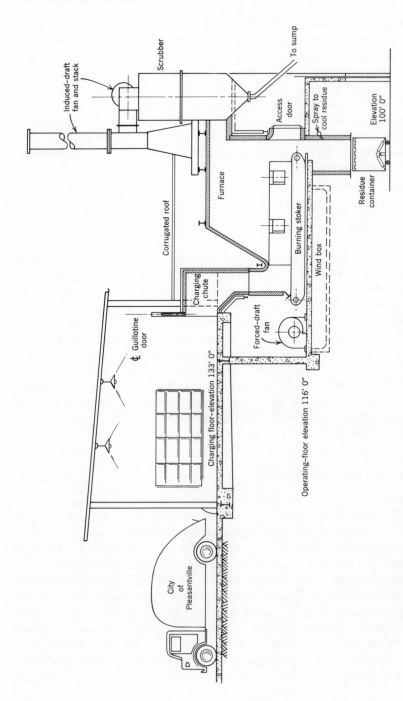

Figure 6-4 Hillside location of floor-dump incinerator (longitudinal section), with typical elevations.

168

scrubbers for the incineration process. These advantages, including heat balances and other engineering data, are covered in detail in recent literature [5, 1]. Reference 1 discusses in depth the difference and similarities of practices in the United States and Europe; these are largely dictated by political and financial considerations, especially as to the availability and cost of fossil fuels.

Figure 6–5 shows a refuse-incineration, sewage-treatment plant. Many such plants have been installed in communities ranging in population from 6000 to 3,600,000. The incinerator gas temperature of 1600°F or more ensures complete reduction of noxious odors, and suitable dust collectors remove the fly ash from the stack.

Combined Facilities for Heat and Power

Combining heat recovery and utilization with incineration in a well-designed and operated continuous-feed plant may be justified. One ton of present-day high-heat-value refuse is equivalent to 70 gal of fuel oil, or four-tenths of a ton of coal in generating steam. Performance data can be readily calculated from the refuse heating values in Table 6–1 and the approximate efficiencies given in Figure 6–6. Many of the following possible uses for the heat developed by incineration of refuse have been proved by operating installations in the United States and abroad:

1. Plant or building heating and cooling by reversible steam cycles with low- or medium-pressure steam.

2. Power generation for plant use comprising high-pressure boilers and electric generating equipment. For heating and cooling steam may be bled off at suitable pressure.

3. Steam and/or power generation for sale to nearby users, such as building, market, or hospital complexes.

4. Sewage-sludge or effluent sterilization by means of flash dryers, vacuum filters, and similar means. In some cases the filter cake may be used as a means of removing the fly ash.

5. Water purification or desalination through evaporators and other processes now available.

6. High-temperature, high-pressure water generation for heating adjacent complexes.

7. Gas-turbine power generation with closed systems so that the fly-ash content of the gas does not have to be removed until the latter has been cooled.

8. Hot-air heaters, provided special alloys are used that will withstand the relatively high metal temperatures involved.

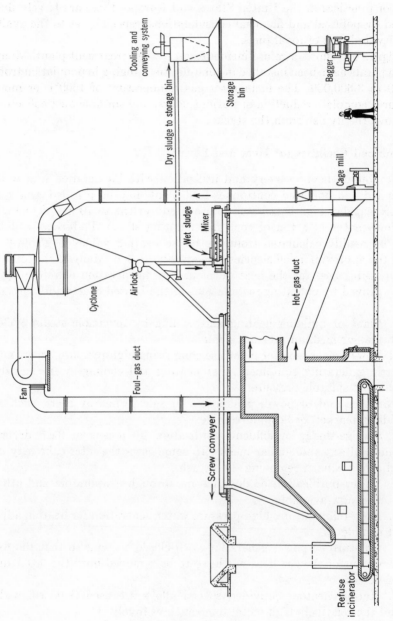

Figure 6-5 Combined refuse-incineration and sewage-treatment plant.

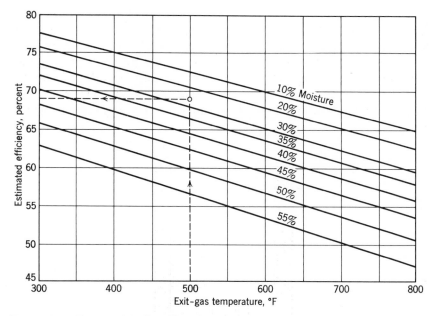

Figure 6–6 Estimated boiler efficiency for refuse incineration. Based on cellulose refuse and 50 percent excess air. Reduce by 3.3 points for 100 percent excess air.

Capacity and Operating Period

Individual incinerator capacities range from about 60 to almost 700 tons/day, with many in the 250-tons/day size in satisfactory operation. Figures 6–7 through 6–10 illustrate typical foreign applications. Figure 6–11 shows a proposed design for the United States (see also Figures 6–17, 6–18, and 6–19). The capacity is usually rated as tons of refuse burned per 24-hr period. To be more meaningful this should be broken down to an hourly rate; for example, a furnace rated at 300 tons/day should burn 100 tons in 8 hr and 200 tons in 16 hr. Therefore the hours of operation must be determined.

This incinerator-operating period—whether one, two, or three shifts per day—vitally affects design factors such as furnace size, as well as labor and capital costs. Most combustion processes provide their greatest output and optimum performance when operated at a steady, continuous rate. Fuel-air ratio, furnace temperature, and cleaning schedules can then be stabilized. In a survey of engineering practices covering municipal incinerators, the United States Public Health Service showed that the median of operational and maintenance costs reported for 8-, 16-, and

24-hr burning periods were $1.40, $1.00, and $0.48 per ton per furnace per day, respectively [2]. The longer burning periods therefore resulted in considerable savings per ton of refuse incinerated. The incoming trucks generally arrive between 8 a.m. and 4 p.m., but adequate storage and reclaiming facilities will allow two- or three-shift operation. A recent tabulation of 64 modern plants by the author showed that 85 percent were designed to operate for two or three shifts per day and for 5 to 6 days per week.

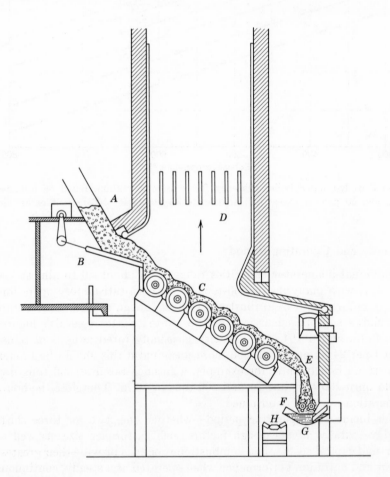

Figure 6–7 Düsseldorf incinerator with rotating-drum grate in a water-wall steam generator. *A* and *B*. Refuse charging system. *C*. Roller grates. *D*. Flue gases to steam generator. *E*. Residue discharge. *F*, *G* and *H*. Residue quenching and handling.

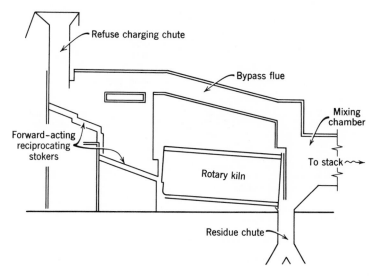

Figure 6–8 Incinerator with Volund reciprocating grate stoker combined with a rotary kiln.

INCINERATOR DESIGN

Facilities for Handling Refuse

The method of handling the refuse after it reaches the plant and the provision for its storage between the time that it is delivered and fed to the furnaces largely determines the burning period. Storage bins must be much larger than would be required for other more familiar combustible materials because municipal refuse is bulky (weighing from 250 to 350 lb/yd³, or 9.25 to 13 lb/ft³) compared with coal (which averages about 50 lb/ft³). Also, refuse does not flow readily out of a large bin or hopper; it tends to cling and mat together, standing almost vertically when compacted in a pile, because of its steep angle of repose. For this reason small incinerators are frequently designed for the delivery trucks to discharge the refuse onto the floor above the furnace, from which it is subsequently pushed, either manually or by a bulldozer, into openings over the furnace (Figure 6–4). This arrangement provides very little storage space. The refuse must be burned as fast as it is delivered to avoid delaying other trucks waiting to be unloaded. Since the waiting trucks represent loss in capacity and cost, many such plants are designed with excess furnace capacity to take care of the peak loads when several

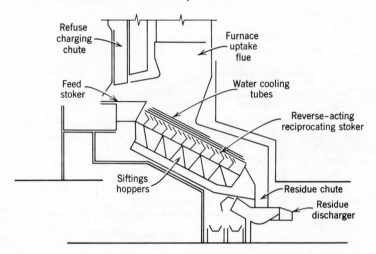

Figure 6–9 Marlin reverse-acting reciprocating grate stoker in water-wall steam-generating incinerator.

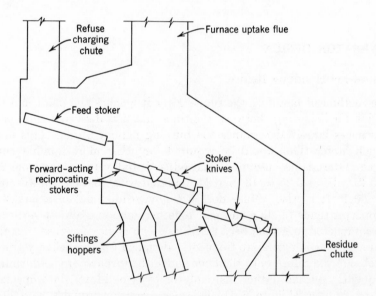

Figure 6–10 Incinerator with a Von Roll rocking grate.

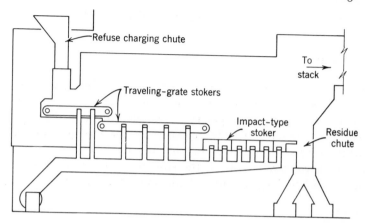

Figure 6-11 Incinerator with a double traveling-grate stoker combined with an impact grate for improved residue burnout.

trucks arrive one after another. A typical hourly load curve of deliveries for 8 a.m. to 4 p.m. pickup (Figure 6–12) shows the materially higher load requirements of the floor dump as compared with the storage bin and crane facility as well as the effect of holidays on subsequent deliveries. Holidays and some weekdays show much greater accumulation of refuse than other days (see Figure 6–2). The lack of storage capacity also makes the burning of all refuse the same day it is delivered imperative to avoid fire hazards, odors, rodent problems, and scattering of the refuse by the wind. If the furnaces are not ample for the peak loads,

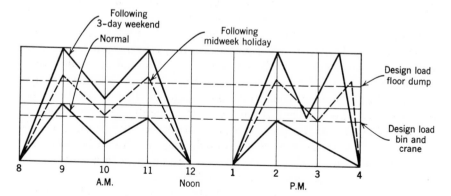

Figure 6-12 Hourly refuse deliveries under various conditions for a typical day.

the refuse may be pushed through half-burned and result in an undesirable residue.

Storage Bin

The installation of a storage bin or pit eliminates some of these problems. The peak loads can be apportioned, and the furnaces can then be sized for considerably less capacity to handle only the average hourly load. However, removal of the refuse from the storage area requires equipment especially designed for the low bulk density and arching tendency of normal refuse. Both the storage capacity and the handling equipment must be large because of this low bulk density.

Cranes

The equipment most widely used in handling refuse is a bridge or monorail crane, connected to a clam-shell bucket or grapple. With this combination the refuse can be picked up from various parts of the bin and moved to other sections for mixing or for clearing the dumping area— or it can be carried along to the charging hoppers as desired. The bridge crane is the more costly but also the more versatile type. With it the bin can be of any desired width, although for practical reasons this is usually limited to 30 ft. Vibrating hoppers, as well as conveyors of various types, have been tried with questionable success. Bin width is limited by the fact that trucks usually discharge their refuse on only one side of the bin, and the angle of flow of refuse is quite steep. If the bin is wide, a crane must keep moving the rubbish as it accumulates from the loading to the far side of the bin. Another fact is that the bridge for the crane must, of course, span the bin with something over on both sides for bucket travel. Excessively wide bins with long bridges are therefore uneconomical.

The monorail crane can move in one direction only; that is, parallel to the center line of the bin or rail. Since it has no bridge, its initial cost is considerably lower; but its range is limited, especially as regards the width of the storage bin. If the bin is too wide, the bucket cannot reach refuse deposited along the sides, where it may stand up almost straight because of its tendency to mat and cling together. In practice the width of the bin is limited to 2 or 3 ft more than the overall width of the bucket, opened wide so that its pointed tines can scrape the refuse down the sides of the bin as the bucket is lowered. The bin should be located at one end and in line with the several charging hoppers to keep the track straight. Generally a curved track is undesirable.

The choice of type of crane frequently depends on the plant layout, but these limiting factors should be kept in mind. Figures 6–13 and 6–14

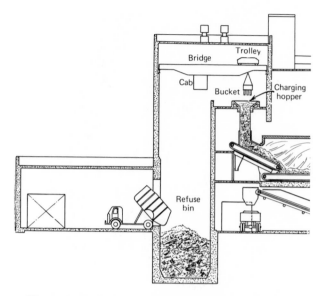

Figure 6–13 Bridge crane for municipal incinerator.

show typical bridge and monorail-crane construction. The need for a spare crane is debatable; money tied up in this surplus item may be put to better use elsewhere. With a well-trained crew and a reasonable stock of spare parts, most routine repairs such as replacing worn cables and bucket parts can be done during weekend or overnight shutdowns. A good program in preventive care is highly desirable. Also it is wise to provide for a work platform and storage of a spare bucket at one end of the bin.

Furnaces

Size

An important factor in the cost of the plant is the number and size of the furnaces installed. In the past individual furnace capacities were limited to about 150 tons/day, so that a 500- or 600-ton plant would require four such units. During the past few years unit capacities have been gradually increasing, especially since the advent of the continuous-feed types; today two 250- or 300-ton units can provide such a capacity, with a marked reduction in first cost, labor, and maintenance. Units of 500 tons/day and more are being proposed.

Some designers have preferred to install several small furnaces—with the thought that if one should break down and require time out for

Figure 6–14 Monorail crane serving two incinerators.

repairs, the remaining units could carry on with the job. There was a time when this theory was widely held in the power-plant field, but experience has shown that there is less chance of trouble with a few large units that are well designed and properly operated than with a number of carelessly maintained small ones. A good maintenance program is the best assurance of reliable performance. The usual weekend shutdown period should provide plenty of time for normal repairs, provided parts are available and the work is properly scheduled and supervised. For similar equipment the cost of two large units might be only two-thirds that of four smaller units with corresponding savings in labor and maintenance. One operator is generally required for each furnace, more or less regardless of its size, and repair costs are likely to be in proportion to the number of units involved.

Furnace sizing, as well as design, affects performance. When the furnace is too small, combustion cannot be completed until some of the volatile combustible matter has passed over the bridge wall, because of the lack of the time factor noted in the section on "Heat Balance." The heat-up time for such a furnace can actually be longer than for a larger unit because the heat generated by delayed combustion in the secondary chamber does not do any good in the furnace. When high-moisture refuse is burned this can reduce the incinerating capacity; this is because a large amount of heat is required for evaporating the moisture, which must be done in the furnace. The steam tables show that it takes 1810 Btu to heat 1 lb of water from 70 to 1600°F, or 30 percent of the 6000 Btu available. At about 70 percent moisture content refuse ceases to be autocombustible; auxiliary fuel must be used to maintain furnace temperature. This is illustrated in Figure 6–15 for a refractory-lined furnace. At no excess air the furnace temperature is only 1900°F; at 50 percent excess air it has dropped to about 1600°F, which is below the desired minimum for complete combustion.

The effect of an oversize furnace is easily shown by assuming that the refuse burned is cut in half, with a similar reduction in combustion air. The radiation, leakage, and unaccounted losses will remain unchanged; whereas the heat loss in the combustion products will be reduced (especially if the excess air is also lowered) so that a substantial heat balance is obtained and the desired temperature is maintained.

Configuration

The incinerator designer has two general types of furnaces to choose from; namely, the long-used batch-feed design in which the refuse is dropped periodically onto the fire from a small hopper—and the continuous-feed type in which the refuse is removed continuously from a much larger hopper by means of the moving grate surface below it. The crane operator keeps the hopper full.

In general batch-feed furnaces are lower in first cost and unit capacity; they are most suitable for plants burning less than, say, 150 to 200 tons/(day)(unit). They come in two types, with either circular or rectangular grates. Each type has a form of grate mechanism that is designed to keep the refuse agitated and burning actively as well as moving down toward the dump trays from which the residue can be removed as it accumulates. Considerable manual effort is required, especially during cleaning; one operator is generally assigned to each unit. A typical rectangular batch-feed furnace is shown in Figure 6–16.

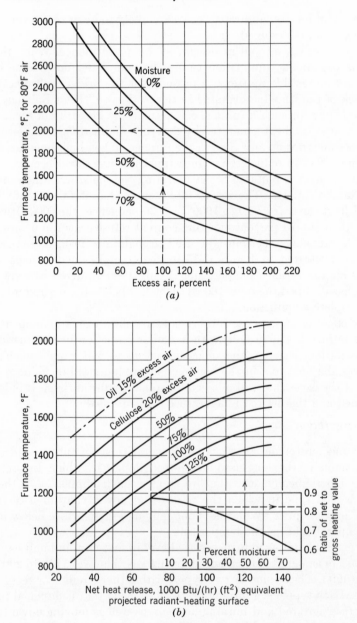

Figure 6–15 Furnace temperatures for refractory-lined (*a*) and water-wall furnace (*b*).

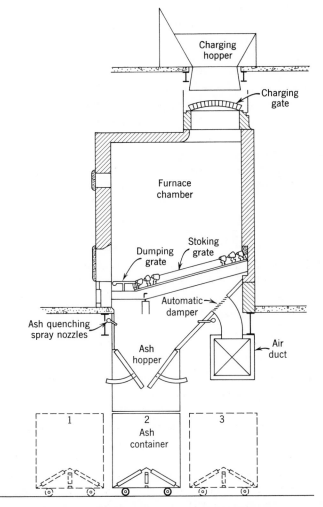

Figure 6–16   Rectangular batch-feed incinerator with a rocking grate.

One major difference in the design of these batch-feed furnaces is the number of charging gates per furnace; this has a definite effect on the performance.

The circular furnace has a single charging gate that is located over the center of the furnace. The furnace is charged at short intervals, the fresh charge falling directly on top of the pile of burning refuse. The refuse forms a cone-shaped pile that gradually burns down with combustion largely on the surface of the pile. Combustion air is supplied through the grate and through the central cone, preferably from forced-draft fans. The refuse is moved slowly down to the dump grates by revolving rabble arms attached to the bottom of the central cone. The residue comprising cans, bottles, ashes, and other noncombustibles is removed periodically at the dump grates. These grates are raised and lowered manually or by power cylinders to reduce labor and speed up the operation materially. When they are lowered, usually one at a time, the operator rakes the accumulated residue down into the ashpit, working around the furnace through four equally spaced cleanout doors.

During the cleaning period the feed-hopper slide gate is closed to keep cold air from entering and refuse from being fed during this interval. Unless the cleanout is done very carefully the incinerating capacity may drop for an appreciable time, depending on the amount of residue on the grate and the skill of the operator. To make up for this lost time the furnace may be forced beyond its normal capacity when refuse feeding is resumed. Cleanout is by far the most troublesome part of the whole operation. As it must be done several times each shift, every effort should be made in designing the plant to facilitate this work by providing adequate operating space, power dump-grate mechanism, and ample furnace capacity.

The rectangular-type incinerator is generally fitted with two or more charging gates and hoppers so that only one section need be inactive during the cleaning period. The remaining sections can continue operation and maintain the furnace temperature at or near normal. This not only improves incineration but also reduces brickwork spalling. The refuse cannot burn properly unless the furnace temperature is kept high enough to ensure rapid drying and burning of the combustibles.

Furnaces of this type are fitted with movable grates that rock or reciprocate in such a way as to carry the burning refuse from the charging end to the dump grate, usually with some help from the operator. These grates are then dumped and the residue is discharged as in the circular furnaces. A typical installation may have a rocking-grate stoker, a hopper for bin-and-crane refuse handling, and several containers for residue disposal. The hopper is designed to hold a grab bucket full of refuse directly on top of the slide gate, which is divided into a right and left half for ease in handling. The crane operator should keep this hopper filled to the floor level at all times.

Continuous-flow hopper-feed incinerators are made in several types, including traveling and chain grates, revolving kilns, rocking or reciprocating grates, rotating drum grates, and combinations of these elements. In one of the most successful designs (Figure 6–17) the traveling grate

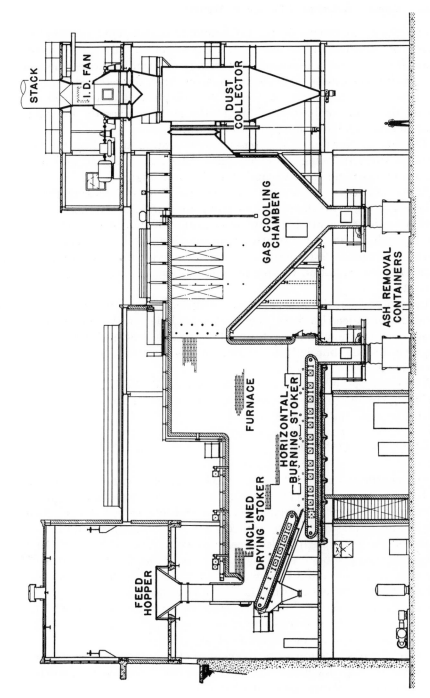

Figure 6-17 Modern continuous-feed, refractory-lined incinerator with traveling-grate stokers.

183

carries the refuse progressively over a number of dampered air compartments, so adjusted that combustion air is admitted only at the points where it is required. The refuse is fed from the bottom of the feed hopper at a continuous rate by a stoker mechanism. First the refuse dries, then it begins to burn on the surface as the volatile matter is driven off; finally the solid combustibles burn as the grate reaches the discharge end, and the residue is dumped into the ashpit.

The city of Newton, Massachusetts, is operating an incinerator that employs three traveling grates in series. The unit, designed and built by the Tynan Incinerator Company, is shown in Figure 6–18.

Many municipal incinerators use reciprocating and rocking grates. The larger sizes are divided from front to rear into several sections, with provision for control of grate movement and addition of combustion air in each section. A recently designed three-section incinerator is shown in Figure 6–19. In these incinerators the refuse moves continuously with sufficient agitation to promote combustion. Manual raking is not needed.

In addition to the traveling and reciprocating grates common in American practice, European incinerators include rotary kilns, reverse-action reciprocating grates, rotating-drum grates, and inclined cones to improve residue burnout, as shown in Figures 6–8, 6–9, and 6–7, respectively.

Various types of mechanical grates have been described in detail, including circular grates with rotating stoker arms, and rocking, reciprocating, impact, and traveling grates [8].

Comparative Performance

The charts in Figure 6–20 show typical capacity variations for one- and four-charging gate batch-feed furnaces, compared with the much more uniform output obtained with the continuous-feed design. Also, the fuel-air ratio is more easily controlled in the continuous-feed furnace; in the batch-feed types it varies because of overfeeding and underfeeding as the charging gates are opened and closed. Furnace temperature is much more uniform, and throughputs are normally higher for the continuous-feed furnace than for the batch-feed designs. The absence of cleaning periods permits sustained operation at the rated capacity for days at a time, and reduces spalling and slagging. For the same reasons less brickwork maintenance is required.

These obvious advantages of the continuous-feed incinerators suggest their use for all capacities, especially above 100 tons/day, and when the optimum in performance is desired.

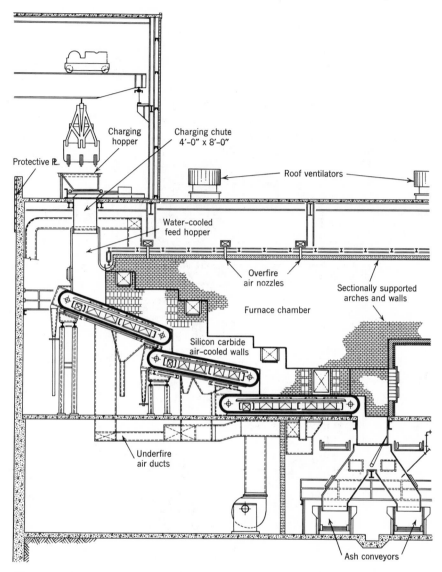

Figure 6–18 Three-stoker incinerator designed and built by the Tynan Incinerator Company.

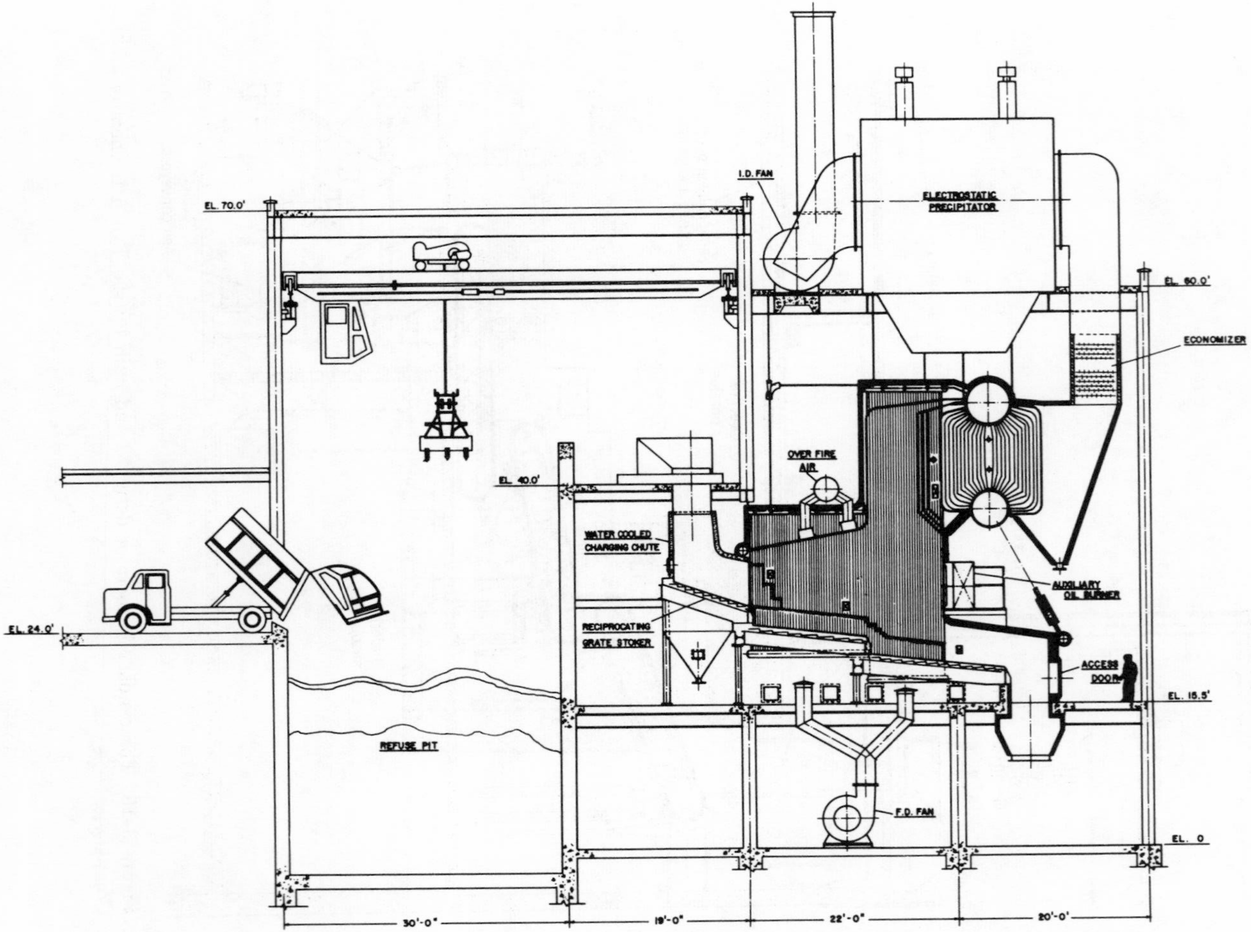

Figure 6-19 Continuous-feed incinerator with rocking-grate stoker.

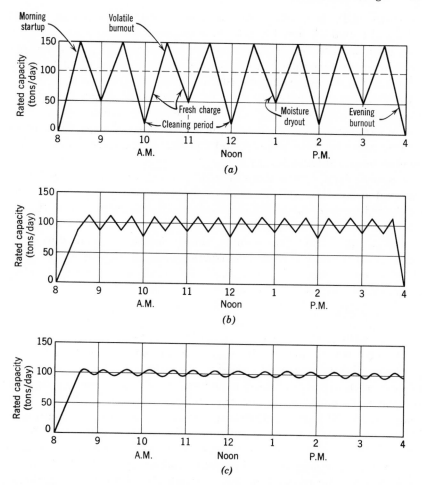

Figure 6–20 Daily operating cycles for three types of batch-feed and continuous-feed incinerators. Batch-feed incinerator with one charging gate—(a); batch-feed incinerator with four charging gates—(b); continuous-hopper-feed incinerator—(c).

Refractory

In modern furnaces refractory repairs can be made on separate portions of the walls or arches because of the sectionally supported air-cooled construction now in common use. The furnace and arch tile are supported on metal clips or hangers, which are bolted to steel buckstays for the side walls and to beams across the top for the arches. The spalled or broken tile can be repaired or replaced without tearing down and rebuilding the entire wall, as is frequently necessary with solid-brick or gravity walls. Individual tile may be replaced in the suspended arches in the same way. What used to be a week's job has become an overnight or weekend chore for plant personnel, with little or no impairment in plant capacity.

Such materials as plastic or castable refractories can be used for the entire furnace or for patching when minor repairs are needed. No longer need fallen arches be a major hazard in the modern incinerator.

The two factors that cause the most trouble and expense in maintenance are refractory slagging and spalling. Slagging is caused by ash buildup on the side walls and arches, frequently because of overloading or excessive flame temperature in some parts of the furnace. Spalling is aggravated by the extreme changes in temperature that occur when batch-feed furnaces are charged or cleaned. Short-time or one-shift operation has been found particularly bad in this respect. As the furnace is heated up quickly in the morning and cooled down again during the night, alternate expansion and contraction of the brickwork causes pieces of refractory to break loose or spall. The more nearly continuous the operation, the less spalling occurs.

Water Walls

Water walls in the furnace have many advantages—as well as certain limitations, which must be thoroughly understood if the desired results are to be obtained without excessive maintenance. The advantages include reduction in refractory maintenance by elimination of exposed furnace walls and arches, and the ability to operate with high flame temperature without excessive slag formation, because the continuous black surface of the water-cooled tubes offers much greater heat absorption. With refractory walls 100 to 200 percent of excess air is required; with water walls only 25 or 30 percent is necessary to obtain the higher temperature. The reduced air flow through the refuse on the grate and the lower gas velocity in the furnace beneficially affect fly-ash pickup and carry-over and materially improve air-pollution control.

However, unless the temperature of the water in the walls is at least 300°F, or well above the dew point of the corrosive gas in the furnace, fireside pitting will result. Also this water must be deaerated and softened in accordance with established boiler procedures; otherwise internal oxygen corrosion and scale formation will contribute to eventual tube failure. From a practical viewpoint preferred conditions are best ensured when a steam or high-pressure hot-water boiler is included in the circulation system and the treated water is continuously recirculated. The saturation, or water temperature, in the steam boiler, 380°F at 200 psi, will ensure adequate metal temperature, which is normally taken as the internal water temperature. This boiler then acts as a heat exchanger, reducing the outlet-gas temperature so that the fly ash can be removed readily by efficient collectors, without additional cooling air or water. Increasing either of these would increase the amount of outlet gas and adversely affect draft losses and the size of equipment required for air-pollution control. Many European and some South American and United States cities have found that incinerators with water walls can be located in the heart of the load area, where hauling problems are reduced and waste-heat utilization is more feasible.

The troubles experienced in the United States with water-wall furnaces have been caused by using untreated water at ambient temperature in the tubes, with once-through flow requiring 100 percent makeup. Internal scaling and corrosion, as well as external pitting, were serious maintenance problems. Boiler applications should be made only by a qualified manufacturer, with knowledge of similar installations where refuse fuels such as bagasse and hogged wood are being burned. Special conditions that must be thoroughly understood and evaluated include (a) tube spacing to avoid fouling in the screen tubes or first pass of the boiler, (b) locating the soot blower, (c) sizing the superheater, and (d) designing the whole furnace to ensure ample flame travel and complete burnout. A recent installation in the United States is shown in Figure 6–19.

Contrary to popular belief, the major function of the walls and roof of the furnace is not to reflect the heat back into the furnace, but to direct the gas flow and prevent the admission of uncontrolled cold air. Even in a refractory furnace the wall temperature is always somewhat cooler than the flame, except perhaps for a few moments after the fire is cleaned or the heat input is otherwise suddenly reduced. Heat flows only from high to low temperature; and, unless the walls are at a higher temperature, they cannot reflect heat back to the flame. This was proved definitely years ago with the advent of water-wall furnaces, many of which now have no refractory next to the fire; yet such furnaces are operated at higher temperatures with much less slagging- or refractory-maintenance trouble and with reduction in the excess air required. Similarly, the heat-up time is affected much less by furnace-wall construction than by the fuel-air ratio during this period. The belief that heat absorption by the cold refractory furnace walls materially delays achieving full furnace temperature also is refuted by experience with water walls. With them heat is absorbed at a much faster rate—possibly 30 times faster than in furnaces without water walls. The wall temperature never gets much above the water temperature of several hundred degrees Fahrenheit, yet in practice the furnace reaches full operating temperature in a comparatively short time. This time interval is usually determined, as an experienced operator knows, by allowable expansion stresses rather than by the firing rate.

In a cold start refractory walls absorb a considerable amount of heat, and it may take several hours before they reach full wall temperature. Calculations show that the refractory in a test furnace wall absorbs about 8,600,000 Btu, but it may take 10 hr or more for this heat loss from the furnace to occur, during which time the excess-air flow, and resultant heat loss in the combustion products, may be correspondingly reduced.

Radiation losses do not start until the outside-wall temperature has risen appreciably above that of the ambient air; thus such losses are negligible during the warm-up period.

Controls and Instrumentation

Combustion-control equipment, such as temperature recorders, and draft gages should be of the best quality for long life and reliability. Thermostatic controls can hold the furnace temperature in continuous-feed incinerators quite uniform by regulating the excess air when variations may occur too fast for the operator to correct. Either the carbon dioxide or the preferred oxygen recorders can determine the excess air and thus aid the operator in regulating the rate of charging combustion air.

Draft gages in the various wind boxes, air ducts, furnaces, and flues indicate abnormal conditions (such as burned out or overloaded fuel beds and fouled-up gas passages) and abnormal furnace draft. They should be of the multipointer type, fully enclosed for cleanliness, and with large, easily read scales. The range of the scales should be carefully selected so that the pointer will normally be near the center, with any deviation quickly observable. The choice between indicators and recorders is suggested in Figure 6–21, which gives typical requirements for a continuous-feed incinerator. In general good power-plant practice should be the guide in selecting the elements for such a control setup.

Dust Collectors

With adequate furnace volume the gas velocity can be kept low enough for particles to remain in the combustion zone and burn to ash. The fine fly ash that does pass out of the furnace should be trapped by a well-designed and adequate collection system. The smoke is best eliminated in the furnace.

Water-sprayed baffle chambers may reduce fly-ash emission to below many local-code requirements. The spray water cools and cleans the gas; however, water consumption is high because the baffle must be kept wet

Draft Gages	*Number per Furnace*
Stoker compartments—one pointer for each zone; total-draft range: −0.5 to +3.0 in. water gage	
Forced-draft-fan outlet duct—draft range: 0 to +4 in.	1
Sidewall-nozzle air duct—one on each side; draft range: 0 to +4 in.	2
Roof-nozzle air duct—one side only; draft range: 0 to +4 in.	1
Furnace draft at furnace outlet—in sidewall; draft range: −0.5 to +.1 in.	1
Differential gage for dust-collector draft loss—draft range: 0 to 6 in.	1
Total number of draft-gage pointers	1

Recorders and Controllers

Furnace temperature—at furnace outlet in sidewall; temperature range: 100 to 2500°F	1
Dust-collector-inlet temperature—temperature range: 100 to 1000°F	1
Draft-control—automatic furnace-draft controller in one sidewall at furnace outlet; draft range: 0.00 to −0.40 in. water	1
Furnace-outlet temperature controller—regulate total air from forced-draft fan to maintain temperature in 1600 to 2000°F range	1
Dust-collector-inlet temperature controller—adjust sprays to maintain 650 to 750°F; include alarm signal for excess temperature above 800°F	1
Emergency sprays at dust-collector inlet—solenoid valve in pipe connected to roof tank in case of normal water-pressure failure	1
Emergency fan-disconnect relay—stop forced- or induced-draft fan, or both, in event of excessive gas temperature to dust collector; interlock forced- and induced-draft fan motors to stop forced-draft fan if induced-draft fan stops	1
Carbon dioxide or oxygen recorder—at furnace outlet in sidewall at rear of furnace	1

Figure 6–21 Suggested instruments and controls for typical continuous feed incinerator.

at all times. Much of this water can be recirculated after the solids have settled out. In some cases suspended baffles have cost more than other types of collectors; in a few instances maintenance has been a problem.

Dry tubular or cyclone collectors can handle gas at temperatures up to about 700°F; therefore the entering gas must be precooled with air or water sprays. When water is used in well-designed fog nozzles it is converted to dry superheated vapor so that a maximum amount of heat is absorbed from the gas and the water consumption is brought within reasonable limits. The water consumption of dust collectors has been evaluated in the literature [3].

Many of the prevailing dry-settling and expansion chambers have proved inadequate; they should not be included in modern installations. The experience of boiler-plant designers can well be followed in this respect. The various types in use or worth consideration are listed below roughly in the order of their relative collection efficiency, from low to high performance, as shown in each case.

1. *Expansion or settling chambers* depend on low gas velocity and gravity settling of the dust particles. When baffles are installed the gas may be made to flow over and under or horizontally around them, with the dust settling in the bottom of the chamber for subsequent removal. To avoid high maintenance in the stack lining the entering gas should be air cooled to less than 1000°F. Collection efficiency is 10 to 20 percent.

2. *Dry-baffled or inertia-separation* chambers cause the gas to reverse direction around or under baffles so that centrifugal force and gravity combine to effect dust removal. Collection efficiency is 30 to 40 percent.

3. *In wet baffled or wet-bottom chambers* the dust-laden gas impinges on the wetted baffle surface, and the dust is carried down into suitable hoppers or flumes. The gas is cooled by sprays. Depending on the baffle and spray arrangement, and on the amount of water used, collection efficiency may vary from 60 to 80 percent.

4. *In mechanical or cyclone-type dust collectors* the gas may be first cooled by air or water to temperatures not over 700°F and then cleaned by centrifugal action of the spinning gas stream in the cyclone. Cooling the gas with a heat exchanger is preferred. Collection efficiency is usually guaranteed at 80 to 85 percent.

5. *Electrostatic precipitators* use an electrical discharge that causes the dust particles to adhere to parallel plates or concentric surfaces. Periodically they are shaken or flushed loose and deposited in hoppers for subsequent removal. Generally this method is not satisfactory for fly ash with carbon content over 25 percent; therefore it is not normally recommended for incinerators, except as a follow-up to mechanical types, or with gas that has been cooled to 700°F or less by passing it through a boiler or other heat exchanger. Collection efficiency may be 95 percent or higher.

6. *Wet scrubbers* designed and installed by commercial organizations have steel shells with suitable baffles and sprays or venturi tubes through which the dust-laden gas is drawn by an induced-draft fan. With these water consumption is relatively high and settling basins are frequently

used, with the overflow recirculated. Water treatment may be required to prevent corrosion. Also, material such as stainless steel or special coatings may be used in the parts exposed to corrosion. A rather heavy stack plume is sometimes characteristic of scrubber applications. This does not pollute the air but may prove unsightly. Reheating the gas that leaves the scrubber to 500°F or so may reduce this plume to acceptable limits. Collection efficiency is generally 95 percent or higher, depending on particle size.

7. *Bag filters* made of silicone-coated glass fibers or similar material have high collection efficiency with dry gas at not over 600°F. Since moisture from cooling sprays clogs the bags, such filters are practical only when the hot gas is cooled by passing it through a heat exchanger such as a boiler. Collection efficiency may be 99 percent or higher.

8. *Packed towers* contain suitably shaped ceramic inserts. In flowing through them the dust-laden gas comes in intimate contact with the wash water, so that the dust is scrubbed out and discharged with the water. Collection efficiency is in the high nineties. Performance with incinerator emissions is subject to tests.

Draft losses in the above dust collectors range up from 0.5 in. water. In general the draft loss is higher when the collection efficiency is higher. With practically all types induced-draft fans are required for satisfactory performance; also, to avoid distortion of rotating parts, the temperature of the gas leaving the collector must not exceed 600°F. Because the draft loss varies as the square of the flow rate, and the fan horsepower increases directly as the gas volume and draft or static pressure, careful attention must be given to excess-air control and gas cooling.

Odor and smoke are best controlled by oxidizing the effluent to inert form such as carbon dioxide and water. To accomplish this the combustion temperature must be at least 1400°F either in the furnace or in an afterburner. Most of the dust collectors listed are ineffective in odor and smoke removal except at a fan-power cost greater than is economical. Chapter 3 covers the principles of gas cleaning in detail.

INCINERATOR OPERATION

Sorting Refuse

The suggestion has been made that noncombustibles such as cans, bottles, metal containers, and tramp iron be removed before the refuse is fed to the furnace. However, experience has shown that these materials help to keep the fuel bed open and porous so that the combustion air can pass through more readily. The same applies to crates, cartons, and other rather bulky rubbish, which burn more readily when they are not flattened or compressed. Shredders or hammer hogs have been tried in some cases but have been discarded for the above reasons. Their best use is with tree branches, Christmas trees, and similar metal-free trash.

Sometimes there is salvage value in the cans and tramp metal, which may be picked out of the residue by magnetic pulleys or cranes, or in unburned paper, rags, and cans for their tin, zinc, solder, and iron content; the salvage market is so variable, however, that the plant should be designed to incinerate all of the refuse collected except that which is too large to pass through the furnace or is hazardous (such as oil drums or similar containers that may contain explosive substances). Increased use of direct oxygen converters in the steel mills has reduced the value of ferrous scrap except for export, and the low cost of producing many paper products from wood pulp has reduced the demand for scrap paper.

Obviously, items such as bed springs, metal refrigerators, hot water boilers, and baby carriages should not be delivered to the furnaces; these should be collected separately and disposed of directly to the dump, when not salvageable. Crushing or macerating can reduce their bulk. The well-designed incinerator should have charging hoppers or grates that will take material up to about 4 ft in its maximum dimension. Undersize hoppers will be an endless source of trouble because of fouling and inability to pass much of the delivered refuse.

High-moisture material such as garbage, discarded fruits and vegetables, and animal carcasses cannot be incinerated without help from more combustible fuel. An ample proportion of dry rubbish—such as paper, cartons, crates, and similar combustibles—will keep the furnace gas temperature high enough for the rapid drying that is the primary reaction and make the use of auxiliary oil burners or other heat sources generally unnecessary. The crane operator can help materially by watching for the discharge of such garbage into the bin and moving it to the top of a pile of drier rubbish before loading it into the charging hoppers. With a properly designed ramp and dumping layout trucks can unload as soon as they arrive while the crane operator keeps the near side or one end of the bin clear for this purpose. This handling tends to mix the refuse, and blend the wet and dry ingredients for easier combustion. Also, with the floor-dump system the bulldozer should do as much mixing as possible to avoid smothering the furnace with a load of wet garbage or market refuse.

Combustion

When the gate of a movable-grate furnace is opened the refuse drops onto the partly burned out fuel bed. The slide gate is then closed and the hopper is refilled, ready for the next charge. As the refuse burns down it is moved along by the action of the rocking grates until it reaches the dumping section, where final burnout is completed.

The temperature in the furnace must at all times be maintained high enough to decompose the refuse and ensure that all the combustion products have been heated above the ignition temperature of the several combustible ingredients involved. This should be at least 1400°F gas temperature at the furnace outlet to ensure complete oxidation of noxious odors, burnout of suspended carbon to avoid smoke, and complete combustion of carbon monoxide to carbon dioxide.

Overfire Air

Each type of furnace has its own requirements, but in general approximately one-half of the total air should be blown in above the fuel bed in the combustion chamber. Air supplied under the grate tends to pick up and carry over light particles of paper and trash, thus increasing dust loading to an undesirable degree. Fly-ash carry-over has been found to increase approximately as the square of the velocity of the gas leaving the fuel bed. Air introduced above the fire helps burn out the combustible gas and reduce excessive furnace temperature but does not increase the dust loading if the fuel bed is not disturbed by jet penetration. Overfire air, introduced above rather than through the fuel bed, is essential for combustion of the highly volatile refuse now encountered. To be effective this air must be supplied by forced-draft fans with sufficient velocity to create turbulence and proper mixing with the combustible gas. Figure 6–22 can help in selecting nozzles and fan capacities. The curves, based on tests by Battelle Memorial Institute some years ago, show penetration of air jets at various diameters and pressures.

Dust-Loading Correction for Excess Air

The stoichiometric air is of course 100 percent of the required combustion air, and the excess air is added to this to determine the total air supplied; for example, for 100 percent excess air the total air is 100 plus 100 or 200 percent of the stoichiometric requirement. Therefore, when correcting dust loading actually measured at 100 percent excess air with that at 50 percent excess air for use in the ASME code, the measured

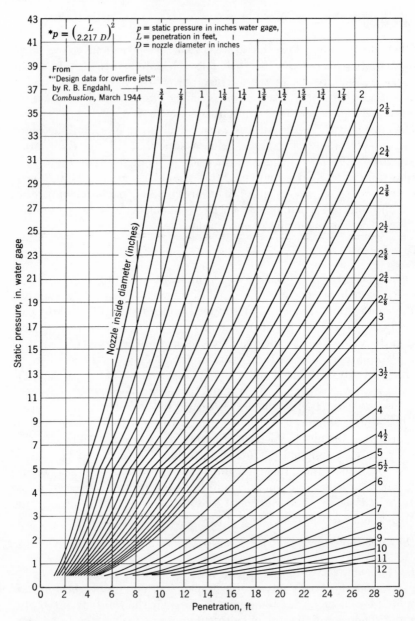

Figure 6–22 Penetration curves for overfire-air nozzles. Nozzle velocity of 1000 fpm. (Prepared by Combustion Engineering, Inc.)

dust loading must be multiplied by the ratio of 200 over 150, which equals 1.33.

The excess air is normally determined from Orsat analyses of the combustion products, the formula being

$$\text{excess air} = \frac{100 \times O_2}{(0.264 \ N_2) - O_2} \text{ percent.}$$

The correction to 50 percent excess air was based on coal in common use at that time, which was frequently burned with 50 percent excess air, or about 12 percent carbon dioxide. Carbon monoxide is assumed to be zero in this formula.

The excess air dilutes the combustion products so that the dust loading per 1000 lb of gas is less at higher percentages, for which reason the correction factor is necessary for comparative results. Since 50 percent excess air is not necessarily equal to 12 percent carbon dioxide for refuse burning, it is incorrect to base the correction on the latter factor—as is sometimes done in test reports.

Excess Air

The maximum furnace temperature in any combustion process is of course obtained when only theoretical or stoichiometric air is supplied. In the case of most fuels, including refuse, this temperature is well above 3000°F, which greatly exceeds the allowable value for practically all solid fuels because of ash fusion and clinkering.

Excess air is air supplied for combustion in excess of that theoretically required for complete oxidation. For years it has been used to control directly the temperature of the furnace exit gas. For most incinerators with the all-refractory furnaces normally used this has required 100 to 200 percent excess air, substantially less when water cooling is provided. Complete combustion of cellulose fuels such as refuse can be accomplished with 50 percent or less excess air when water walls have been installed. There is therefore plenty of leeway in using excess air to cool the furnace, which should be done only as required to prevent excessive slagging and to reduce furnace maintenance.

The radiation loss seldom averages more than 1 to 3 percent of total heat input and depends largely on furnace-wall construction and temperature differential between furnace and ambient air. Heat absorbed in the brickwork of the furnace before stable conditions are obtained is a function of the weight of brick involved, temperature rise, and specific heat of the brick. It seldom exceeds 5 percent even when operation starts with a cold furnace.

All fuel—whether solid, liquid, or gaseous—actually burns largely as a gas, the heat from the furnace walls serving to dry out the moisture and to volatilize the combustible matter from the fuel bed. Both of these reactions are endothermic, or heat absorbing. The gas then combines with the overfire air and burns out to carbon dioxide and sulfur dioxide. These reactions are exothermic, or heat producing. The fixed carbon burns out in the fuel bed, producing additional heat. Combustion in the fuel bed is therefore from the top down, except for the bottom layer of carbon, next to the grate surface.

The hottest part of any furnace is in or near the top of the flame where combustion to carbon dioxide has been completed. The flame itself is therefore the hottest arch possible and should be permitted to burn out directly above the refuse on the grate to ensure rapid drying and ignition. When a pound of carbon burns only to carbon monoxide it liberates 4450 Btu, or about one-third of its total heat; when the carbon burns to carbon dioxide it liberates all of its total heat.

Furnace size has a rather secondary effect on the temperature obtained, since it is possible to obtain practically the same flame temperature over a 2:1 or more range in heat-release rates; that is, for a given burning capacity the furnace size can vary quite widely without material effect on the temperature, provided the excess air is carefully controlled.

Heat Balance

There are only three ways for the heat generated by the burning refuse to escape from the furnace:

1. Some will be carried away in the combustion products, which include the air admitted under and over the fuel bed, air admitted through leakage, combustible matter in the refuse, and moisture in the air and refuse.

2. Some of the heat will be absorbed in heating the brickwork during the initial start-up and carried away by radiation from the outside of the furnace when equilibrium has been established.

3. The remainder will be carried away as sensible heat in the hot residue that drops into the ashpit and as unburned combustible matter either in the gas stream or in the residue.

In a heat balance all of the heat input must be accounted for, with an item at the end of the tabulation covering any minor differences between the calculated losses and the total heat input from the burning refuse. When there is no excess of heat input over losses the furnace temperature will remain uniform (Figure 6–23), which occurs with 200 percent excess air. When the excess air is reduced to 150 percent there is an excess

Refuse burned per hour (lb)	1000
Gas temperature in furnace (°F)	1500
Ambient temperature (°F)	80

Excess air at furnace outlet (percent)	200		150	
Heat losses:	Total Btu	Percent	Total Btu	Percent
In combustion products	5,430,000	90.5	4,300,000	88.3
By radiation through walls[b]	156,000	2.6	156,000	3.2
By leakage of air through walls	180,000	3.0	180,000	3.7
From unburned combustible and unaccounted for	234,000	3.9	234,000	4.8
Total heat losses	6,000,000	100.0	4,870,000	100.0
Total heat input (1000 × 6000)	6,000,000		6,000,000	

At 200 percent excess air the furnace temperature remains steady because input equals losses.

At 150 percent excess air the furnace temperature rises to 1700°F because heat input is greater than losses; additional excess air will have to be supplied.

[a] Assumed to have 20 percent moisture and 6000 Btu/lb as fired.

[b] Furnace walls composed of 9 in. of firebrick and 8 in. of red brick.

Figure 6–23 Heat-balance calculations for a refractory furnace with 200 and 150 percent excess air to show effect on furnace temperature.

of input over losses, and the temperature rises, as noted. Should the excess air be increased to 250 percent, the total loss would increase to 6,920,000 Btu compared with the input of 6,000,000 Btu, with a consequent drop in temperature.

The radiation loss of 2.6 percent can be reduced to perhaps 1.5 percent by using an insulated setting or increased to 4 or 5 percent by reducing the thickness of the setting. In this case the outside-wall temperature would be so high that the operators would be endangered. The close relationship between furnace construction and outside-wall temperature is clearly illustrated in Table 6–2.

The above heat balances show that by far the greatest (but also the most easily controlled) heat loss is due to air flow into and through the furnace. Changing the radiation loss would require rebuilding the furnace. The loss due to moisture is determined by the refuse itself, which generally must be taken as it comes. Hence ensuring the proper furnace temperature for good incineration depends on the fuel-air ratio or the amount of combustion air admitted to the furnace in a given time.

It is interesting to note that the moisture loss in combustion products in above example is only about 6 percent as compared with 84 percent for the dry gas loss; thus moisture control is not only difficult but relatively unimportant compared with total air control.

Residue Disposal

More attention should be given to equipment for handling and removing residue than it has received in the past. With improved furnaces and better combustion equipment and controls there should be very little unburned combustible matter in the residue. From combined collections the residue will average about 0.65 yd³, or 520 lb per ton of refuse fired on the dry basis—a burndown ratio of approximately 4:1 in weight and 10:1 or better in volume.

Table 6–2   Effect of Furnace-Wall Construction on Cold-Face Temperature and Radiation Loss in Still Air at 1600°F Furnace Temperature

Wall Construction	Cold-Face Temperature (°F)	Radiation Loss [Btu/(ft²)(hr)]	Percent
4½-in. plastic refractory	440	1356	6.5
9-in. plastic refractory	319	730	3.5
9-in. firebrick;			
8-in. red brick	274	545	2.6
4½-in. plastic refractory;			
1-in. insulation	262	495	2.4
8-in. plastic refractory;			
4-in. insulation	153	150	0.72
Continuous water walls;			
2½-in. insulation	140	138	0.66

The simplest method in common use for handling residue from batch-feed furnaces is the hopper with quenching sprays and a slide gate. The hot residue is cooled by the sprays as it drops from the grate into storage, where it is kept until it is discharged by gravity into a truck, conveyor, container, or other disposal equipment. With the container system the residue may be accumulated and stored overnight. Figure 6–9 shows a multiple-container design.

In the rectangular incinerator the residue is discharged into a hopper that is fitted with adjustable bottom plates, so designed that the residue can be deposited into any one of three or more sealed containers that are mounted on tracks and can be rolled into position by hand. As each container is filled it is moved along and replaced by an empty one. The filled containers may be picked up by truck and carried to the dump during the day shift or left in place overnight and removed to the dump the next morning. This arrangement permits two- or three-shift operation of the furnaces, with residue hauling during the day shift only. Should all containers become filled, the bottom plates may be closed and the hopper used for storage. Properly placed sprays quench the residue as it is discharged.

The residue must be dumped several times each shift, depending on the burning rate and amount of noncombustible matter in the refuse. There is some loss in capacity during this period.

Conveyor systems frequently installed in the larger plants with continuous feed grates are designed so that the hot residue will drop into a water-filled sump that has a conveyor at the bottom Figure 6–24. This conveyor carries the wet residue up a trough inclined to ensure considerable dewatering of the residue and then dumps it into trucks or containers. The water level in the sump is maintained high enough to form an air seal at the discharge end of the grate and thus avoid undesirable leakage into the furnace.

Other forms of conveyors, also hydraulic sluices with dewatering tanks or ponds, have been used with satisfactory results—the choice depending on the engineers preference, the local terrain, and the disposal facilities.

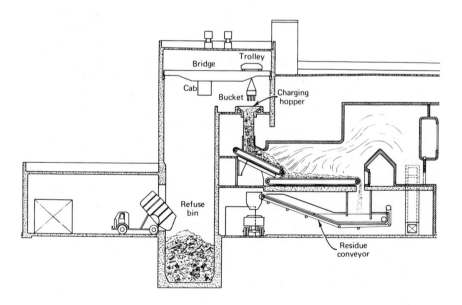

Figure 6–24 Residue disposal from continuous-feed furnace by inclined, water-sealed conveyor.

CALCULATIONS

Combustion

As many experienced engineers are finding out, the apparent variations in refuse characteristics for any given plant location are more an illusion than a reality. A well-known western university reports [6]:

Analyses of the samples of the refuse burned during the tests showed that the heat of combustion on the moisture- and ash-free basis was very close to that of cellulose—8000 Btu per lb—regardless of the chemical composition of the fuel. This fact simplified the calculations on the heat process and made it possible to predict the behavior of the material when it was burned. The most important variable was the moisture in the refuse, which is considered as the burden, or the energy consuming load on the incinerator.

It is true that the moisture content varies from day to day, depending on the refuse-pickup and handling system and the weather; however, proper control of the excess air can largely control this variation. Furnace temperature need not vary so widely that incineration capacity or efficiency is unduly affected. Auxiliary fuel, such as gas or oil, should be needed only for heating up after a weekend shutdown or for maintaining the heat output when boilers or other heat exchangers are included.

Converting the weight of refuse to be fired into British thermal units (Btu), the heat unit that has long been used by power-plant and other furnace designers, has greatly simplified calculations for furnace determinations. Instead of selecting the grate and furnaces on the weight units or the number of tons or pounds to be burned per hour, the refuse is analyzed for heating value or compared with known material, and its net Btu value is then multiplied by the pounds to be fired per hour to obtain the total heat input, which is usually expressed in million Btu per hour.

Combustion engineers have known for years that the air requirements for complete combustion of most solid fuels are quite uniform on the as-fired Btu basis; many of the variables such as moisture and noncombustible matter having relatively little effect, except on the gas weight and furnace-cleaning or residue-discharge functions. With this concept established the air requirements and combustion products for the most usual operating conditions can be calculated and plotted, as shown in Figure 6–25.

The major design factors are then developed from the above starting point as shown in the following example. Many of these data can be

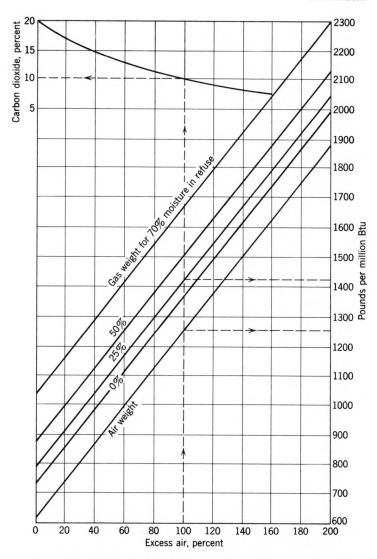

Figure 6–25 Combustion air and gas weights for various excess-air values and moisture contents of municipal refuse.

tabulated for the various incinerator sizes in common use or expressed in curve form, whichever is most convenient. This procedure follows methods used in designing furnaces for many standard fuels—such as coal, hogged wood, and bagasse—as well as for liquid and gaseous fuels such as oil and natural gas. Performance guarantees have been made on this basis for years in all of these categories.

EXAMPLE. Combustion Calculations (Basis: 100 tons/day)

1. Heating value (moisture- and ash-free basis)   8000 Btu/lb.
2. Moisture   25 percent.
3. Ash and other noncombustible matter   12.5 percent.
4. As-fired heating value, 8000[1 − (0.25 + 0.125)]   5000 Btu/lb.
5. Firing rate—100 tons/day (100 × 2000/24)   8340 lb/hr.
6. Firing rate—heat input (5000 × 8340)   41,720,000 Btu/hr.
7. Excess air required according to Figure 6–15 to burn   100 percent.
   refuse at 2000°F flame temperature
8. Theoretical combustion-air requirement based on heating   5 lb.
   value of 8000 Btu per pound of cellulose
9. Total air weight per pound of combustible (theoretical   10 lb.
   air + excess air)
10. Total air weight per million Btu (1,000,000 × 10/8000)   1250 lb.
11. Weight of water per pound of refuse [0.25/(1 − 0.25)]   0.333 lb.
12. Weight of water per pound of combustible [0.333/(1.0 −   0.38 lb.
    0.125)]
13. Total gas weight per pound of combustible equals air   11.38 lb.
    weight + water weight + 1 lb combustible
14. Total gas weight per million Btu of combustible   1420 lb.
    (1,000,000 × 11.38/8000)
15. Grate area, assuming heat-release rate at 300,000   139 ft².
    Btu/(ft²)(hr)(41,720,000/300,000)
16. Combustion rate (300,000/5000)   60 lb/(ft²)(hr).
17. Furnace volume, assuming heat-release rate of 20,000   2085 ft³.
    Btu/(ft³)(hr)(41,720,000/20,000)
18. Furnace height (2085/139) or (300,000/20,000)   15 ft.
19. Total air weight, 100 percent excess air (1250 × 41.7)   52,100 lb/hr.
20. Air volume, including furnace leakage at 100°F as calculated below.
21. Weight of air at 100°F, atmospheric pressure is 0.0709   12,300 ft³/min.
    lb/ft³ (52,100/0.0709 × 60)
22. Air supplied by forced-draft and overfire fans at 85 percent   10,450 ft³/min.
    of above (15 percent leakage air assumed)
23. Gas weight leaving furnace (1420 × 41.7)   59,400 lb/hr.
24. Gas volume at 2000°F corresponding to 59,400 lb/hr   61,200 ft³/min.
25. Vertical gas velocity in furnace above fuel bed   440 ft/min.
    (61,200/139)
26. Horizontal gas velocity at furnace outlet for 8-ft furnace   510 ft/min.
    width (61,200/8 × 15)
27. Cross-sectional area in duct or chimney at gas velocity of   30.6 ft².
    2000 ft/min (61,200/2000)

These values are easily converted to the weight basis, such as per ton of rated capacity or similar units for comparison. Start by moving the decimal point in the above example two places to the left. Thus 1 ton/day equals 83.4 lb/hr, equivalent to 417,000 Btu/hr.

The combustion air becomes 521 lb/(hr)(ton of rated capacity). The gas weight at 100 percent excess air is 594 lb/(hr)(ton). Equivalent weight at 50 percent excess air is 467 lb/(hr)(ton), or 1110 lb/ (million Btu)(hr). The combustion rate is 300,000 ÷ 5000 = 60 pounds per square foot per hour. The furnace volume becomes 417,000 ÷ 20,000 = 21 ft³/ton.

The dust-loading limitation set by the ASME model code of 0.85 lb of fly ash per 1000 lb of gas at 50 percent excess air converts to 0.85 × 1110 ÷ 1000 = 0.943 lb per million Btu. Other code requirements can of course be converted to suit.

Fan sizes are obtained by taking the above air and gas volumes, together with static pressures based on test data, and adding the standard fan tolerances, in accordance with established power-plant practice.

It will be noted that variations in the unit heat value, which affect all other results, are taken care of by correcting for moisture and ash. For an as-fired 4000-Btu refuse with, for instance, 50 percent moisture plus ash content the total heat input becomes 33.4 million Btu/hr. For industrial refuse, which may average 6000 Btu/lb or more as fired, this heat input becomes 50.0 million Btu/hr. The design should therefore be based on the highest average unit heat value that may be expected, and the resultant million Btu total is that used in subsequent calculations.

Specific Heat

In the calculation of any heat transfer the specific heats or heat-absorbing capacities of the various media involved are constantly being used, so a brief discussion of this factor is desirable. Specific heat is the Btu required to raise the temperature of a unit weight, usually 1 lb, of a substance 1°F. Every medium—whether solid, liquid, or gas—has its specific heat values, which vary with the temperature and pressure. These values are compared with that of water, which is one, or unity, all others being expressed as a decimal. Engineering handbooks have tables of specific heats of various materials, and the curves in Figure 6–26 show those in common use in incinerator design. Mean specific heat is the average above a specified base, such as 80°F; instantaneous specific heat is the specific heat at the particular temperature involved. For our purpose the specific heat may be considered as the ability of the various media to absorb and carry away the heat generated, which heat may be transferred later to some other medium, such as cooling air, water sprays,

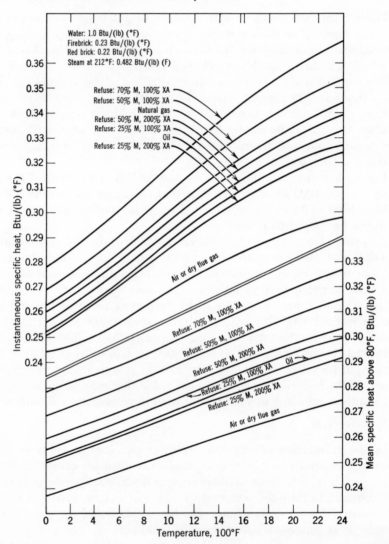

Figure 6–26 Specific heats of interest in incineration. Abbreviations: M-moisture; XA-excess air.

furnace walls, or a heat exchanger such as a boiler. Heat is never lost or dissipated. Water, with its specific heat of unity, will pick up and carry away four times as many Btu per pound up to 212°F as will air with its specific heat of about 0.25. The water then absorbs its heat of vaporization, or 970.4 Btu/lb, as it turns to steam at atmospheric pressure, or a total of 1112 Btu above an ambient temperature of 70°F, compared with 35.5 Btu/lb for air. Water is therefore the most efficient medium for gas cooling, whether in the form of sprays or in a boiler.

Furnace Temperature

The three "Ts" of combustion—namely, *time, temperature,* and *turbulence*—are so closely related that they must be considered together in determining the performance of an incinerator. Without sufficient time for the combustible matter to burn the desired temperature will not be obtained, and without adequate turbulence neither the time nor temperature requirements will be achieved.

For any given furnace and fuel there is a fairly definite and rather easily determined optimum furnace temperature that can be calculated when certain controlling factors are known or assumed. Such calculations have been used for many years in the design of boiler furnaces, but their application in incinerator design is not so well known or understood.

Controlling factors include fuel characteristics such as ash-fusion temperature and moisture content for solid fuels, also unit heating value, as well as characteristics such as furnace size and design, water-cooled or refractory-wall construction, including boilers or other heat-absorbing, or "black," surfaces. The S/V ratio (heat-absorbing surface versus furnace volume) is sometimes used to express the latter factor.

The principles involved are the same for any fuel. The moisture content in a solid fuel such as refuse cannot be readily measured or controlled. However, the total air flow is quite easily regulated; it is the major factor in carrying away the heat generated and in controlling the furnace temperature, as the attached heat-balance calculations and charts for a typical incinerator designed to burn municipal refuse illustrate. Flame-temperature calculations for a typical refuse fuel are given below; they are plotted for various excess-air and moisture combinations in Figure 6–15.

The designer who has had experience with power-plant layouts will recognize the close similarity between boiler furnaces and incinerators, especially in the need for close fuel-air ratio control and optimum performance in the furnace to ensure overall satisfaction. Although there is still much to be learned in the art of incinerator design, the science of materials and their proper use in design, construction, and plant opera-

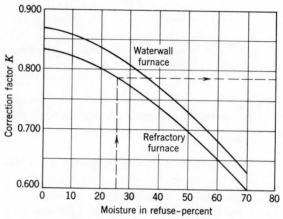

Figure 6–27   Correction factors for net versus gross heat input.

tion and control is developing. By proper coordination of available knowledge and experience much uncertainty of the possible performance can be eliminated. Firm guarantees covering overall operation are possible.

The skill of the plant personnel is still a major factor. Proper training of those in charge, adequate supervision, and careful maintenance are required.

Calculation of Flame or Furnace
Temperature for Refuse Incinerator

Refractory Furnace

Refuse:

Heat value (Btu/lb as fired)		6000

Analysis (percent by weight):

Proximate		Ultimate	
Moisture	20	Carbon	50.0
Noncombustible	5	Hydrogen	6.0
		Oxygen	40.0
		Nitrogen	3.0
		Sulfur	1.0

Excess air at flame port (percent)	200
Combustion products (lb per million Btu heat input)(gross)	2040
Ambient air temperature t (°F)	80

Furnace temperature T (°F) $= \dfrac{\text{Btu input (net)}}{W_g \times Sp_m} + 80°\text{F}$,

where Btu input (net) = Btu input (gross) $\times$ correction factor K (Figure 6-27),

$\qquad\qquad W_g$ = combustion products in pounds per million Btu,

$\qquad\qquad Sp_m$ = mean specific heat above 80°F,

$\qquad\qquad (T - t)$ = gas-temperature rise above ambient temperature in °F,

$W_g \times (T - t) \times Sp_m$ = Btu input (gross) $\times K$,

$$(T - t) = \frac{\text{Btu input (gross)} \times K}{W_g \times Sp_m}.$$

For above conditions

$$(T - t) = \frac{1{,}000{,}000 \times 0.805}{2040 \times 0.28} = 1410°\text{F},$$

$$T = 1410 + 80 = 1490°\text{F}.$$

REFERENCES

[1] Eberhardt, H., "European Practice in Refuse and Sewage Sludge Disposal by Incineration," *Proc. ASME 1966 Incinerator Conference*, New York, May 1966, pp. 124–143.

[2] Eberhardt, H., "Municipal Incinerator Design—A Survey of Engineering Practices," *Proc. ASCE, J. of the Sanitary Engineering Div.*, **84**, SA3, 1677 (June 1958).

[3] Fife, J. A., and R. H. Boyer, Jr., "What Price Air Pollution Control," ibid., pp. 89–96.

[4] Kaiser, E. R., "Chemical Analyses of Refuse Components," *ASME Paper No. 65WA/PID-9*, November 1965.

[5] Meissner, H. G., and E. R. Shequine, "Refuse Incinerators," Chapter 31, *Combustion Engineering*, Combustion Engineering, Inc., New York, 1966.

[6] *Municipal Incinerators*, Technical Bulletin No. 6, University of California, Berkeley, November, 1951.

[7] Wegman, L. S., "Planning a New Incinerator," *Proc. ASME 1966 National Incinerator Conference*, New York, May 1964, pp. 1–7.

[8] Westergaard, V., H. G. Meissner, and W. T. Clark, "Types of Mechanical Grates for Incinerators," *ASME Paper No. 62-WA-259*, November 1962.

7

ON-SITE INCINERATION
OF SPECIAL INDUSTRIAL WASTES

Richard B. Engdahl*

In the constant effort to reduce the costs of industrial production there is naturally a continuing effort to find ways to recycle the process wastes into the process itself. Most industries have characteristic patterns of reprocessing of their own wastes, although in many situations it is cheaper to obtain more raw material than to attempt to reprocess partially processed raw material. There is need, however, for constant review of the feasibility of reuse, because changing technology can render feasible practices that may for a long time have seemed unfeasible. Nevertheless, almost always there is a certain residue from most processes that cannot be reused and must be disposed of. In a few cases neighboring industries are able to use the waste products. Where such alternative uses cannot be economically justified or where the residues are for some reason hazardous to store or to transport, disposal becomes necessary. Combustion or incineration is often the most logical method of destruction of many of these wastes.

---

* Fellow, Mechanical Engineering Department, Battelle Memorial Institute, Columbus, Ohio.

TYPES OF COMBUSTIBLE WASTES

Tables 7–1 and 7–2 list some industrial wastes compiled by Hescheles [16].

SPECIAL INDUSTRIAL WASTES

Solid Industrial Wastes

The worthless, solid, combustible residues from manufacturing processes are sometimes the most difficult of all wastes to dispose of because of their heterogeneity, ash or moisture content, and poor or hazardous burning properties. In a few cases the incinerated residue may have some value that may help to pay for the difficult incineration process [9]. Otherwise it must finally be deposited at some further cost in a sanitary landfill. The following examples illustrate the problems involved and methods for solving them.

Figure 7–1 shows the simplest and most common type of incinerator used for commercial and industrial wastes when the amount of waste to be disposed of is no more than several hundred pounds per hour. This is known as a Class III incinerator according to the Incinerator Institute of America.

The sloping refractory hearth is intended as a drying surface to hold the freshly charged refuse while it dries by heat radiation from the flame and adjacent hot refractory. Usually the dried refuse quickly ignites and burns by surface burning with air supplied from the ashpit to the primary chamber through a cast-iron grate. In some designs the hearth is replaced by a sloping grate. The slope of the hearth or grate helps the operator to push the burning mass down to the lower grate to make way for a fresh charge. Although its simplicity has made it popular, its satisfactory performance in a nuisance-free manner depends to a very great degree on the care and skill of the operator. If the incinerator becomes overloaded with a wet, slow-burning mass, overfire fuel burners and secondary burners may be used in the so-called settling chamber to speed up the drying and help prevent incomplete combustion in the exhaust gases. Particularly when the primary attention of plant-maintenance personnel is given to other tasks, the batch-fed incinerator is too often overcharged at infrequent intervals. If then to speed up burning they manually poke the charged mass with a firing tool, they are likely to produce excessive amounts of fly ash, charred paper, and smoke.*

For many dry, low-ash wastes, such as scrap wood in moderate quantities, this simple batch incinerator is satisfactory if skillfully used. With

---

* See Figure 5–14 for instructions on operating small incinerators.

Table 7-1 Waste Fuel Analysis

Type of Waste	Heating Value (Btu/lb)	Volatiles (percent)	Moisture (percent)	Ash (percent)	Flash Point (COC°F)	Fire Point (COC°F)	Sulfur (percent)	Dry Combustible (percent)	Density (lb/ft³)
Coated fabric—rubber	10,996	81.20	1.04	21.20	265	270	0.79	78.80	23.9
Royalite	20,299	81.90	0.37	9.62	270	280	0.04	90.38	24.3
Coated felt—vinyl	11,054	80.87	1.50	11.39	165	170	0.80	88.61	10.7
Coated fabric—vinyl	8,899	81.06	1.48	6.33	155	175	0.02	93.67	10.1
Ensolite—expanded	10,216	88.90	0.35	9.99	265	300	0.32	90.01	5.7
Missile—rubber scrap	12,238	71.36	1.69	24.94	340	360	1.17	75.06	28.9
Fuel-cell spray booth	12,325	79.10	1.74	20.67	125	130	0.25	79.33	9.5
Missile—rubber dust	9,761	62.36	0.87	36.42	250	260	1.06	63.58	9.9
Banbury—rubber scrap	13,242	60.51	1.74	4.18	145	180	0.53	95.82	34.9
Polyethylene film	19,161	99.02	0.15	1.49	180	200	0	98.51	5.7
Foam—cloth backed	10,185	55.06	0.37	31.87	215	235	1.44	68.13	11.3
Uppers—cloth	7,301	70.73	1.25	24.08	295	300	0.40	75.92	13.5
Foam—scrap	12,283	75.73	9.72	25.30	185	240	1.41	74.70	9.1
Tape—resin-covered glass	7,907	15.08	0.51	56.73	300	330	0.02	43.27	9.5
Fabric—nylon	13,202	100.00	1.72	0.13	625	640	0	99.87	6.4
Fuel-cell bladder and tire cord	15,227	87.35	1.24	3.57	270	290	0.55	96.43	19.5
Vinyl scrap	11,428	75.06	0.56	4.56	155	165	0.02	95.44	23.4
Liquid waste	13,140	100.00	3.2	1.04	68	68	0.07	95.76	53.0

Test for volatiles was run independently. The difficulty of getting two similar samples accounts for the higher percentage of volatiles.

Table 7-2 Waste Fuel Analysis

Type of Waste	Heating Value (Btu/lb)	Volatiles (percent)	Moisture (percent)	Ash (percent)	Sulfur (percent)	Ignition Temperature (°F)	Ash-Fusion Temperature (°F)	Density Specific Gravity
Batch stock (raw)	14,177	60.29	0.46	13.36	1.07	220	2200	1.112
Uncured frictioned duck	9,343	76.44	2.57	9.16	0.52	200	2580	0.628
Cured duck	{16,454	69.94	1.47	3.76	1.28	280	2220	0.957
	{11,306	78.08	1.47	11.95	0.86	260	2270	1.004
Wire-braid hose	{12,846	52.13	0.40	11.67	0.87	280	2480	2.109
	{7,820	68.83	2.18	15.30	0.29	220	2500	2.047
Uncured C.I.	9,842	55.76	1.73	24.25	1.62	300	2360	1.415
Cured C.I.	10,150	49.89	1.66	20.31	0.60	320	2240	1.478
Cured matting trim	7,228	51.13	0.99	41.47	0.84	300	2240	1.593
Packing trim and C. I.	10,413	53.77	0.99	27.79	0.92	300	2260	1.430
Cured flash and molded goods	15,442	62.38	0.91	9.13	0.97	320	2800	1.232
Miscellaneous yarn (rayon and cotton)	7,759	90.28	6.59	0.61	0.15	220		0.504
Semi-cured tubes	15,392	49.70	1.24	2.03	0.70	320	2720	1.279
Miscellaneous tubing and die strip	11,352	64.15	0.49	30.54	2.23	300	2120	1.209
Rubber tape	9,629	54.51	0.54	31.63	0.56	280	2460	1.698
Ensolex trim	9,139	65.22	0.40	22.99	0.72	260	2800	0.179
Miscellaneous rubber	11,335	59.42	0.37	23.37	1.50	260	2240	1.252

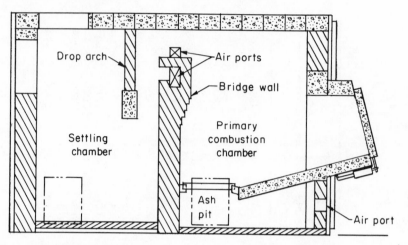

Figure 7-1 Simple industrial or commercial incinerator, Class IIIA, Incinerator Institute of America.

large amounts of wet or high-ash materials, however, some method of mechanical firing becomes necessary. Figure 7–2 shows one common design of a mechanically fired unit [14]. Thus the charging door can be superseded by a mechanically operated sliding door leading to a charging hopper. The hearth may be replaced by a sloping mechanical grate that uniformly moves the burning charge downward toward the ashpit. Over-

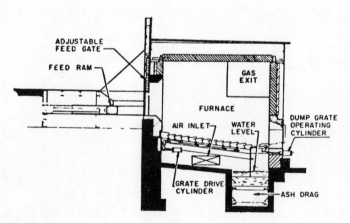

Figure 7–2 Longitudinal section through industrial incinerator with reciprocating grates.

fire-air jets of high-velocity air should provide intense mixing in the primary chamber but must not be permitted to impinge on the burning charge unless the charge happens to have such low ash content that no appreciable amount of ash will be lifted by the high-velocity air. Again, to some extent the satisfactory operation depends on the skill and care of the operator.

Figure 7–3 shows a common type of top-charged, cylindrical incinerator that uses a slowly rotating cast-iron cone with rabble arms to provide some agitation on the grate. Ash is discharged at the periphery of the grate. One industrial plant preferred to use this type of furnace with the rabble arms removed [22]. A growing disadvantage of this method of top charging is that each time a charge of dusty refuse is dropped into the furnace a cloud of dust is carried away by the hot gases, causing momentary overload of any dust-collecting device.

If large quantities of dry, finely divided wood or paper wastes are supplied to such refractory furnaces and if an ample quantity of air is present, the temperatures in the combustion chamber can very quickly exceed the safe operating limit for the refractory walls and deteriorate them [13].

Where the waste is extremely wet a preliminary drying chamber may be necessary; however, if the fuel can be burned in suspension, it can be burned satisfactorily without preliminary drying.

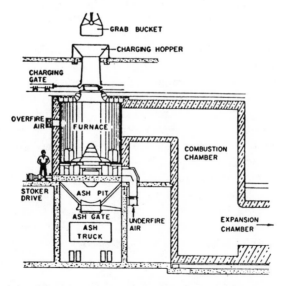

Figure 7–3 Longitudinal section through cylindrical batch-feed furnace.

Some highly variable industrial solid wastes are being successfully incinerated in refractory-lined rotary kilns [10]. The rotating action helps to average out any effects of variability in feed rate or charge composition. Auxiliary-fuel firing can assist in drying wet wastes. A major disadvantage is that the cascading action of the burning refuse as the kiln rotates gives all fine ash in the charge ample opportunity to become entrained in the exhaust gases. Usually good dust collection is required to meet increasingly restrictive air-pollution-control ordinances.

Wood Wastes

The largest quantity of industrial refuse that must be disposed of is probably wood waste. In most woodworking establishments the waste is usually in the form of chips or small pieces that can be conveyed through elaborate pneumatic conveying systems to a woodburning boiler for steam or power generation [5]. Usually such disposal installations must be supplied with some auxiliary fuel to maintain a steady supply of steam or power at times when the waste-wood supply fluctuates. In many other installations, where the process is not primarily woodworking, the wood accumulates from shipping and packing operations. One method of disposal of such material is by hauling it to a landfill or municipal incinerator, but, because of the hauling costs in many areas, on-site incineration may be more feasible. In the past single-chamber, multiple-chamber, and other types of burners have been used. Such simple units can be quite satisfactory in many localities if they are carefully fed with the waste fuel, supplied with sufficient air and turbulence, and not overloaded.* However, in densely populated areas they may emit too much flyash and finely divided charcoal from the stack for them to be acceptable to the neighbors unless good dust-collection equipment is applied. As a practical matter, the difficulties of feeding a nonuniform waste fuel at a satisfactorily low rate is a matter of more skill than the workmen in industrial-waste disposal usually possess. Accordingly, too often waste-burning installations emit considerable smoke because of overloading, lack of air, and insufficient mixing.

The economics of waste-wood utilization has been shifting appreciably in recent years; hence there is less and less wood for incineration. Engdahl [12] points out that in many areas where sawmills can feed sawdust and chips into available boxcars conveniently, these can be moved economically at regular intervals to nearby papermills, where the waste wood can be used in papermaking.

---

* See Chapter 5 for details on construction and operation of single- and multiple-chamber incinerators.

Conical Burner

In lumber areas and at many industrial installations the so-called conical burner for wood waste has been a common sight. Increasingly, however, as concern with air pollution has increased, the conical burner has been subject to considerable criticism, and in many areas it has been replaced. The simple burner is basically a bonfire with a very large conical steel cover standing over it. The top of the cone is usually open, with a hemispherical steel screen to arrest large burning particles. The waste wood is fired through an open firing door at ground level or occasionally through a pneumatic pipe that delivers the fine waste from the plant. Air may be admitted through an opening near the ground level or occasionally through tangentially arranged overfire-air jets; in some cases it may be blown through an air duct into the center of the floor of the conical chamber. The outer shell of the steel structure is kept comparatively cool by the shielding of an inner conical shell that is attached to the outer shell and held some distance away from it by an insulating air space. Because wood normally has a very low ash content, these burners can be operated at a low rate so that unburned charcoal particles are not carried out at the top, and the dust emission from them is low. However, as usually operated, many tests have shown that the cinder loss from conical burners is excessive. Boubel [7] and others [21] have shown from an extended series of tests on conical burners at lumber yards that very few of them would be able to meet today's air-pollution regulations. Again, if equipment can be devised to feed the waste fuel to conical burners at a low rate and in a uniform manner, and if sufficient air and turbulence are supplied to match the fuel rate, the smoke can be negligible and the amount of cinder carry-over can be very low [11]. In populated areas, however, it is likely that some form of dust collection would be required to meet dust-emission regulations.

Silo Burners

Somewhat similar to conical burners, but distinctly more promising, are the silo type of wood burners. These are tall, vertical, cylindrical steel chambers, refractory lined, which provide a high-temperature environment for burning wood waste. Hence potentially they can be operated with a clear discharge. However, the problem with the silo burner often is incomplete combustion caused by excessive fluctuation in firing rate.

Wide variations in temperature result in excessive emissions of smoke, large dust particles, and hydrocarbons. Furthermore, large fluctuations in temperature substantially reduce life of the refractory lining [18].

The Bay Area Air-Pollution-Control District has found that silos can be modified to operate satisfactorily as described by Johnson in 1961 [18]:

Silos which are in satisfactory condition or readily repaired may still be useful if the following modifications are added:

Alternate No. 1(a)   Add a "Dutch Oven" which converts the silo into a multiple chamber incinerator.

(b)   Install a storage bin and controlled feed system from the bin to the ignition chamber of the modified burner.

Alternate No. 2   Install bin and feed system as above. Continue to burn in the silo as above.

Experience in Los Angeles has conclusively demonstrated that modifying silo burners by the addition of the "Dutch Oven," which actually converts this unit into the Los Angeles type multiple chamber incinerator, will meet all of the requirements of Regulation 2, providing a satisfactory controlled feed system is installed.

Regulation No. 2 limits smoke to no more than #2 Ringelman for no more than 3 min/hr, limits particulate emissions to 0.2 gr/ft^3 corrected to 6 percent oxygen, and limits hydrocarbons and carbonyls each to 50 ppm.

Incinerators for Automobile Bodies

Processing of scrap automobiles requires that oil, upholstery, insulation, rubber, and other combustible materials be removed from the steel. This is usually done by burning. Open burning creates too much black smoke; hence incinerators have been devised to confine the process [19]. If ample air can be directed to all burning surfaces, combustion is rapid and reasonably complete. The difficulty lies in supplying enough air and turbulence in the inner recesses. Afterburners are often required to completely burn the combustible gases as they enter the incinerator chimney. A recent development is to burn sheared cubes of automobile-body scrap in a rotary kiln [2]. This subject is described in greater detail in Chapter 8.

Liquid Wastes

Many industrial processes discharge large quantities of liquids that are either combustible themselves or carry combustibles; both must be burned to avoid water pollution. Accordingly, these contaminated liquids are frequently passed through settling chambers or flocculators to increase the concentration of solids and from there into filter beds or vacuum filters to remove large quantities of the water. The resulting filter cake is then fed to an incinerator for disposal. Merman et al. [23] have given a detailed description of a sludge-disposal system at a refinery. The concentrated sludge, consisting primarily of oil and oil emulsions with considerable quantities of dirt and grit, is shipped to an incinerator on an endless belt. The filter cake is 21 percent solids, 27 percent oil, 52 percent water, and 15 percent ash. It has a heating value of about 10,000 Btu/lb. A rotary-hearth incinerator 25 ft in diameter is used to burn the filter cake. The incinerator operates at between 1400 and 2000°F, and the waste gases pass through a waste-heat boiler from which considerable useful steam is derived. A detailed breakdown is given on operating costs. It is pointed out that this incinerator has required extremely high maintenance. Primarily, the repair required is to the rabble arms used to agitate the sludge on the rotary hearth.

Another version that has been developed for industrial sludge is the rotary kiln. One example of this is at Eastman Kodak. It consists of a drum that is 70 ft long, 10 ft in diameter, and revolves at 2 rpm [22]. The first portion of the kiln consists of a drying section in which the temperature is kept at 600 to 800°F to dry the sludge. The remaining 50-ft length of kiln is the burning zone.

In both of these sludge burners the heat for drying and ignition is supplied by oil or gas burners. In the case of the rotary kiln it became desirable after some years of operation to replace a scrubber by a tall refractory-lined afterburner heated to 1200°F.

As mentioned earlier, one problem of the rotary kiln, and to some extent of the rotary-hearth furnace as well, is that as the fine, dry, hot materials cascade from one surface to another, the hot-gas stream tends to carry them away in the flue gases.

Papermill Liquid Effluent

Problems of stream pollution from combustible fibers and "white water" from titanium dioxide in papermill effluent have been solved by a fluidized-bed combustion system [28]. The residue can be recycled in the plant process.

Cyanide Wastes

Figure 7–4 shows a batch system for forming a slurry with 1200 lb of cyanide solids from automotive plating processes and 75 gal of water, which is then mixed with 1200 gal of soluble oil [4]. The mixture, sprayed upward from the bottom of a cylindrical furnace and stack, burns at a rate of 150 gal/hr. Until the system reaches an operating temperature of 2500°F the flame is assisted by a natural-gas pilot.

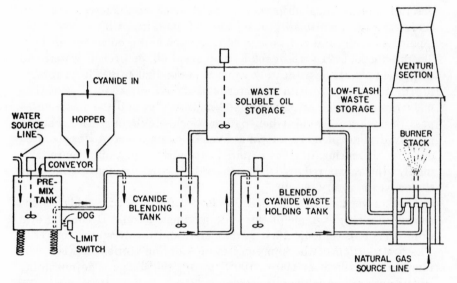

Figure 7–4 Schematic of cyanide-waste-disposal system. Cycle shown disposes of blended soluble oil-cyanide wastes. System design permits use of natural gas to bring burner to temperature and to permit burnoff of either soluble oil waste or low-flash waste, in addition to blended waste.

"Red Water"

The incineration of "red water," a by-product of the explosives industry, has been successfully achieved in a specially built rotating kiln [10]. The composition of this liquid is given below.

Constituent	Percent by Weight
Sodium sulfite	2.9
Sodium nitrite	2.0
Sodium nitrate	4.1
Trinitrotoluene	0.1
Sodium salt of dinitrotoluene	
sulfonate (and other isomers)	8.2
Water	82.7
Total	100.0

This liquid is fired at the rate of 1500 lb/hr into a 30-in.-diameter steel drum, 16 ft long, rotating at 30 rpm. A black sodium carbonate and sodium sulfate ash is continuously scraped from the drum. Fuel-oil consumption is about 5 gal/hr.

Aluminum Chloride Sludge

A refinery-process waste from the manufacture of polybutane is a nonvolatile sludge containing aluminum chloride. An incinerator has been developed to burn this sludge as it is sprayed into the chamber [20]. If the hydrogen chloride or alumina formed exceeds local air-pollution regulations, an exhaust-gas scrubber is needed.

Open-Pit Burner

Monroe [27] has shown that for certain types of wastes, principally those with less than 1 or 2 percent ash, an open pit supplied with high-velocity air jets angled downward across the pit is an effective and nuisance-free burner for solid, liquid, and gaseous wastes. Peskin [29] has described applications of this method, which, if its limitations are recognized, has a place, especially for liquid wastes.

Gaseous Incineration

Some process exhaust gases are not combustible (e.g., sulfur dioxide, nitric oxide, and hydrochloric acid) and hence cannot be controlled through incineration. Other effluents contain appreciable amounts of combustible hydrocarbon gases or vapors, frequently odorous in very low concentrations [24]. Usually these can be incinerated either by direct flame, with or without a flame-supporting fuel, depending on the amount of combustible, or they may be oxidized catalytically.

Whatever incineration of gaseous wastes is attempted, the combustion must be complete. If not, products of partial oxidation, such as aldehydes or organic acids may be discharged; these are more odorous than the original gases [32]. Given below are the temperatures required for odor destruction of some odorous fatty acids that are commonly released by sewage sludge [3, 31]:

Fatty Acid	Temperature (°F)
Acetic acid	1370
Propionic acid	1370
Butyric acid	1425
Caproic acid	1425

The Los Angeles County Air-Pollution-Control District requires [26] that "gases from heated reduction of inedible animal matter (rendering, drying, dehydrating, digesting, cooking, or evaporating. . .) be incinerated at 1200°F for at least 0.3 second. . . ." Other work by that district

[25] indicates that direct-flame afterburners for varnish cookers should be operated at temperatures ranging from 1200 to 1400°F in a chamber with a gas velocity at least 15 fps and a residence time of 0.5 sec.

Direct-Flame Burning of Waste Gases

The most common and effective form of direct-fired gaseous incinerator is a refractory chamber large enough to allow ample residence time for the burning gases, on the order of 0.5 to 1 sec. Intense mixing with high-velocity jets of air is extremely important, particularly with hydrocarbons, to avoid decomposition and smoke formation. Since irregular flow of the waste gas from a process is often a serious source of flame instability and incomplete combustion, gas storage may be required. On the other hand, if the gas is laden with condensible vapors or is hazardous to store, a continuous-flow system is essential. In that case continuous positive ignition by spark or auxiliary flame is the only way to ensure stable burning. When the waste gas is readily ignitable and when the supply is steady, a refractory chamber may not be necessary since the air-cooled gas-turbine combustion chamber provides excellent mixing and burning in a small space.

Economically unrecoverable mixtures of refinery vapors are commonly burned in the open atmosphere, usually by a flare elevated high enough above the surroundings to prevent ground-level nuisance from the products of combustion. Many of these vapors readily form smoke when they are burned unless they are mixed rapidly with air. Smokeless flare designs into which steam or air jets have been incorporated to provide such mixing have been described in detail [1].

Gaseous-Incineration Costs

Examples of the economics of direct-flame incineration for three different industrial odor and fume problems are given in Table 7–3 [6]. Primary heat recovery in this table refers to the use of a heat exchanger to extract heat from the exhaust direct-flame unit in order to preheat the fume entering the unit. Secondary heat recovery refers to the use of a second heat exchanger to extract further heat from the exhaust for use elsewhere in the plant.

Where very lean mixtures of uncontaminated air and gas must be burned, catalytic afterburners may be applicable [8]. Their advantage is that the surface combustion can take place at moderate temperatures and with quite dilute mixtures. Their disadvantage is that the catalyst can become poisoned by atmospheric contaminants; also, inadvertent overheating can cause loss of catalyst.

Table 7–3   Typical Economic Evaluations

	Direct Flame without Heat Recovery	Direct Flame with Primary Heat Recovery	Direct Flame with Secondary Heat Recovery
Sponge-rubber curing oven			
Fuel: #2 fuel oil at 9.4¢/gal			
Operation: 6000 hr/yr			
Fuel cost	$ 26,000	$10,650	
Fuel savings	None	15,950	
Bone-meal process:			
Fuel: natural gas at 40¢/10⁶ Btu			
Operation: 3200 hr/yr			
Incineration-fuel cost	$ 33,400	$ 33,400	
Dryer-fuel cost	18,300	1,300	
Total fuel cost	51,700	34,700	
Total fuel savings	None	17,000	
Glass-fiber curing oven:			
Fuel: natural gas at 56¢/10⁶ Btu			
Operation: 8400 hr/yr			
Incineration-fuel cost	91,000	21,100	$17,600
Oven fuel cost	49,800	49,800	30,400
Total fuel cost	140,800	70,900	48,000
Total fuel savings	None	69,900	92,800

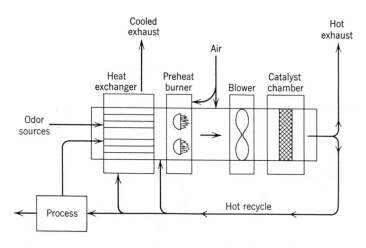

Figure 7–5 Odor control by catalytic combustion [15].

Combustion can elevate the temperature of the gas stream about 55°F per Btu per cubic foot of gas flowing [30]. Thus, if a dilute gas-air mixture has a heat value of only 10 Btu/ft³, the resulting temperature increase of 550°F can sustain self-supporting combustion of some gases once the system is up to temperature. If the gas stream contains enough combustibles to be near the lower explosive limit (LEL), insurance regulations require that air be added to dilute the mixture to less than one-fourth LEL.

Figure 7–5 is a schematic diagram [15] of a catalytic odor-burning system with heat recovery. Table 7–4 [15] shows the estimated cost for this system compared with the costs of a direct-flame system under four different combinations of fume calorific value and degree of heat recovery.

Table 7–4 Estimated Operating Costs of Odor Control by Catalytic Combustion and Direct-Flame Afterburning[a]

Heat Recovery Provided	Calorific Value of Gas	Item of Cost	Cost in $/(hr) (1000 scfm)	
			Catalytic Method	Direct-Flame Method
None	Nil	Fuel	0.51	1.06
		Depreciation	0.09	0.06
		Maintenance	0.09	0.06
		Labor and power	0.03	0.03
		Total	0.72	1.21
None	¼LEL	Fuel	0.03	0.58
		Depreciation	0.09	0.06
		Maintenance	0.09	0.06
		Labor and power	0.03	0.03
		Total	0.24	0.73
One-half of input	Nil	Fuel	0.27	0.56
		Depreciation	0.11	0.10
		Maintenance	0.11	0.10
		Labor and power	0.03	0.03
		Total	0.52	0.79
One-half of input	¼LEL	Fuel	−0.20	0.10
		Depreciation	0.11	0.10
		Maintenance	0.11	0.10
		Labor and power	0.03	0.03
		Total	0.05	0.33

[a] Data from [15].

Waste-Heat Recovery

The sheer waste of heat energy, which is characteristic of incinerators, calls repeated attention to the desirability of waste-heat recovery, especially in industrial plants that often need large amounts of heat for their main processes. Usually a brief analysis of the amount of heat wasted, the proportion of that heat which is realistically recoverable, the irregularity in availability of the heat, and the amortization cost of equipment suited to salvage of the heat leads quickly to the conclusion that the investment required for heat recovery is economically unjustifiable. There are exceptions [17], and as technology and energy costs rise certain recovery opportunities may become feasible.

REFERENCES

[1] American Petroleum Institute, *Manual on Disposal of Refinery Wastes*, Vol. II, 5th ed., New York, 1957.
[2] Anon., "Economical Kiln Process Generates High-Quality Scrap from Auto Bodies Without Air Pollution," *Industrial Heating*, **34**, 666–668 (April 1967).
[3] Anon., "Fume Incineration Effective for Odor Pollution Control," *Industrial Heating*, **33**, 1266–1272 (July 1966).
[4] Anon., "New Way to Incinerate Cyanide Wastes," *Air Engineering*, **4**, 36–37 (1962).
[5] Barkley, J. F., and R. E. Morgan, *Burning Wood Waste for Commercial Heat and Power*, U.S. Bureau of Mines Information Circular 7580 (1950).
[6] Benforado, D. M., C. E. Pauletta, and N. D. Hazzard, "Economics of Heat Recovery in Direct-Flame Incineration," *Air Engineering*, **9**, 28–32 (March 1967).
[7] Boubel, R. W., M. Northcraft, A. Van Vliet, and M. Popovich, *Wood Waste Disposal and Utilization*, Oregon State Engineering Experiment Station Bulletin No. 39 (1958).
[8] Brewer, G. L., "Odor Control for Kettle Cooking," *J. APCA*, **13**, 167–169 (1963).
[9] Butler, F. E., J. D. Moseley, and G. F. Molen, *Recovery of Plutonium from Incinerator Ash*, Dow Chemical Company, Rocky Flats Division, Report to U.S. Atomic Energy Commission, Contract AT(29–1)–1106 (1964).
[10] Challis, J. A., "Three Industrial Incineration Problems," *Proc. 1966 ASME Int. Incinerator Conf.*, pp. 208–218.
[11] Droege, H., H. C. Johnson, L. Clayton, and T. McEwen, "Performance Characteristics and Emission Concentrations from Various Type Incinerators," *Information Bulletin*, Bay Area Air-Pollution-Control District, San Francisco, 1963, pp. 1–63.
[12] Engdahl, R. B., "Process By-Products and Wastes as Fuels," *Pyrodynamics*, **2**, 165–176 (1962).
[13] Engdahl, R. B., and J. D. Sullivan, "Municipal Incinerator Refractories Practice," *ASTM Bulletin* (September, 1959).
[14] Heaney, F. L., "Furnace Configuration," *Proc. 1964 ASME Nat. Incinerator Conf.* pp. 52–57.
[15] Hein, G. M., "Odor Control by Catalytic and High-Temperature Oxidation," *Ann. N.Y. Acad. Sci.*, **116**, 656–662 (1964).

[16] Hescheles, C. A., "Burning Industrial Wastes," *Proc. Metro. Engr. Council Air Resources, Symposium on Incin. of Solid Wastes*, LC Cat. 67–25957, 1967, pp. 60–74.

[17] Hescheles, C. A., "Thermal Recovery Systems from Burning Industrial Wastes," *ASME Publication No. 64–WA/PID–11* (1964).

[18] Johnson, H. C., "Methods and Costs of Wood Waste Disposal in the Bay Area," *Information Bulletin*, Bay Area Air-Pollution-Control District, San Francisco, 1961, pp. 6–61.

[19] Kaiser, E. R., and J. Tolciss, "Smokeless Burning of Automobile Bodies," *J. APCA*, **12**, 64–73 (1962).

[20] Kay, J. B., "Incineration of an Aluminum Chloride Complex," *Industrial Water and Wastes*, **5**, 112–113 (1960).

[21] Kreichelt, T. E., *Air Pollution Aspects of Teepee Burners Used for Disposal of Municipal Refuse*, Environmental Health Series, U.S. Public Health Service Publication No. 999–AP 28, U.S. Department of Health, Education, and Welfare, Washington, D.C., 1966.

[22] Merle, R. L., "Kodak Park Waste Disposal Facilities," *Proc. 1966 ASME Nat. Incinerator Conf.*, pp. 202–207.

[23] Merman, R. G., P. J. Ferrall, and G. T. Foradori, "Sludge Disposal at Philadelphia Refinery," *J. Water Poll. Control Fed.*, **33** (11), 1153–1165 (November 1961).

[24] Meyers, F. D., and J. Waitkus, "Fume Incineration with Combustion Air at Elevated Temperature," *J. APCA*, **16** (7), 378–382 (1966).

[25] Mills, J. L., W. F. Hammond, and R. C. Adrian, "Design of Afterburners for Varnish Cookers," *J. APCA*, **10**, 161 (1960).

[26] Mills, J. L., R. T. Walsh, K. D. Luedtke, and L. K. Smith, "Quantitative Odor Measurement," *J. APCA*, **13**, 467 (1963).

[27] Monroe, E. S., "New Developments in Industrial Incineration," *Proc. 1966 ASME Int. Incinerator Conf.*, pp. 226–230.

[28] Overall, J. E., "Burn Your B.O.D.," *American Paper Industry*, **48**, 51–53 (1966).

[29] Peskin, L., "Open Pit Incinerator," *J. APCA*, **16** (10), 550 (1967).

[30] Ruff, R. J., "Profits from Waste Gases," *Chemical Engineering Prog.*, **53**, 377 (1957).

[31] Sawyer, C. N., and P. A. Kahn, "Temperature Requirements for Odor Destruction in Sludge Incineration," *J. Water Poll. Control Fed.*, **32**, 1274–1278 (1960).

[32] Wallach, A., "Some Data and Observations on Combustion of Gaseous Effluents from Baked Lithograph Coatings," *J. APCA*, **12**, 109–110 (1962).

8

INCINERATION IN METAL SALVAGE

Elmer R. Kaiser

Furnaces of special design are useful in fire cleaning metals of combustibles (such as paint, oil, wood, rubber upholstery, insulation, and brake linings) as well as in melting lead, zinc, and aluminum from iron and steel scrap. They reduce the labor of manual dismantling in metal-salvage operations.

The chapter presents principles and examples for accomplishing these results in a manner that meets modern requirements for air-pollution control.

Cleaning metals by fire has practical advantages in scrap-metal and salvage trades. Often it is the only economical process for salvage. In the past it was common to burn combustibles from scrap automobiles, insulated copper wire, electric motors, oil drums, and the like in the open; but the smoke was excessive, and such open burning has largely been outlawed. Controlled burning in special furnaces developed for the purpose produces a better product and at the same time meets local air-pollution-control standards.

Senior Research Scientist, School of Engineering and Science, New York University, New York, N.Y.

The basic requirements for incineration in metal recovery are the following:

1. Temperatures high enough to maintain gasification and combustion of the combustibles.

2. Temperatures low enough to prevent more than superficial oxidation or loss of the metals being salvaged.

3. Air supply ample for complete combustion.

4. Low consumption of auxiliary fuel.

5. Smokeless chimney effluent and stack gases—low in dust, fly ash, noxious gases, and odors.

6. Throughput sufficient to justify the investment in furnace equipment, fuel, labor, and maintenance.

SMOKELESS BURNING OF AUTOMOBILE BODIES

Fire effectively cleans automobile steel for the scrap trade. The materials to be removed include paint, insulation, rubber, upholstery, plastics, and low-melting-point metals. Although these materials burn in the open, such burning has been outlawed in many areas because of the smoke. The incineration is smokeless only when conducted in a properly designed and operated furnace.

Since about 1958 a number of automobile-body incinerators have been in use in the United States with capacities of up to 50 bodies an hour. A unit capable of consuming 28 bodies in 8 hr and suitable for a small scrap yard has been developed by New York University with grant support by the United States Public Health Service* [2].

Before the auto body is put into the furnace, the motor, radiator, battery, wheels, spare tire, and gasoline tank are removed; the glass is broken; and the trunk cover is opened. The undercarriage may or may not be removed, but the wheels are because tires burn longer than the other combustibles. The bodies are moved into the primary furnace on carts or conveyors. A pint or two of fuel oil—never gasoline—is sprayed on the upholstery to accelerate the ignition. The burning requires about 30 min; it can be speeded up by jets of air blown into the body from sidewall nozzles. Much smoke is generated in the primary chamber because the temperature is deliberately kept too low (1000°F) for complete combustion of the smoke particles (carbon). This smoke is burned either in an afterburner in which the exhaust-gas temperature is raised to 1500°F by an oil or gas burner or burners or in the refractory-lined chimney.

* Plans available at $5.00 prepaid from New York University, attention of the author.

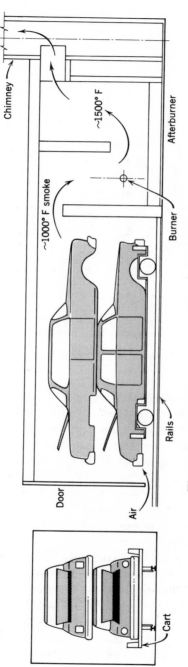

Figure 8-1 Features of small automobile-body burner.

229

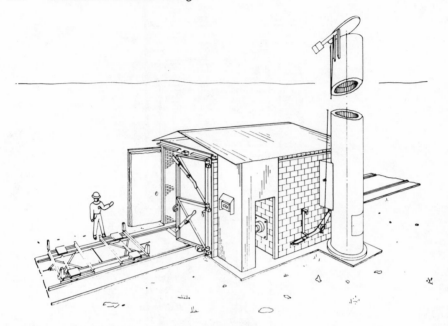

Figure 8–2 Double-end automobile-body burner with afterburner. Capacity: 28 bodies a day.

Figure 8–1 illustrates the principle for a small unit that burns two bodies at a time, one supported on the other. Figure 8–2 shows the exterior of a double-ended furnace for burning two bodies at a time, or 28 per 8-hr day. In operating this unit one cartload of two bodies is burned in the main chamber while the second cart is unloaded and reloaded by crane or fork-lift vehicle. When the bodies have finished burning the doors at both ends of the furnace are opened. The burned load is pushed out by the incoming load, and the process is repeated. While the main doors are open a chimney-cap chamber is kept partly closed to conserve afterburner fuel.

The crane or large fork-lift truck used in transferring the auto bodies can also be used to dunk the bodies in a tank of water to wash off the dust and sluice out the ash, glass, and dirt from the interior of the bodies.

Where a larger capacity is needed twin-cart or conveyor furnaces are used. Figure 8–3 is a diagrammatic section of the conveyor type. With it, the bodies are loaded in front of vestibule A while conveyor B is stopped. Door C is opened, and the conveyor moves until body D has moved inside chamber E. The conveyor is again stopped, and door C

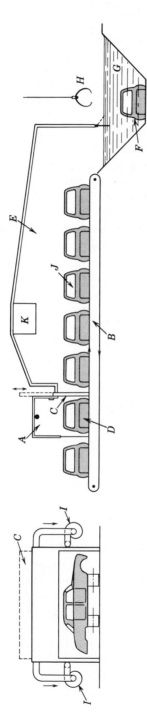

Figure 8-3 Conveyor type of automobile-body burner. *A.* Vestibule. *B.* Conveyor. *C.* Door. *D.* Automobile body. *E.* Chamber. *F.* Burned body. *G.* Tank. *H.* Grapple. *I.* Exhaust blowers. *J.* Jet nozzles. *K.* Gas exit.

is closed. In the meantime burned body F has been discharged off the end of the conveyor into tank G. When it is removed by grapple H the steel is clean and ready for baling, shredding, or shearing. Copper wire, control panels, heaters, and other copper-bearing articles should be removed after the wash and before further processing. While door C is open some smoke escapes from chamber E, but it is caught at the top of vestibule A and promptly returned to chamber E by exhaust blowers I. The blowers discharge into manifolds along the furnace, from which the gases and induced air from A are discharged via jet nozzles J into the burning bodies. Additional air is admitted along the floor of chamber E. The smoky gaseous products of combustion, flowing out of the chamber at K, pass through an afterburner on their way to the stack. Their flow rate is controlled by the cap damper on the chimney so as to prevent smoke leakage from the top of chamber E and to restrict excess infiltration of air. When the system is properly balanced the amount of fuel needed for the afterburner is very low.

The minimum initial cost of the small automobile-body incinerator is about $27,000. The conveyorized units cost about $1200 to $1500 per unit of body capacity in 8 hr; for example, an incinerator burning 13 bodies an hour (100 a day) would probably cost at least $120,000 to $150,000, exclusive of cranes. The oil consumption of the afterburner in the larger units would approximate 5 gal per auto body and 8 to 8.5 gal per body in the small incinerator.

Smokelessness depends on maintaining a temperature of 1450 to 1500°F in the gases leaving the afterburner. By the burnout of the suspended carbon (soot) in the flue gases, the particulate matter in the gases will probably be in the range of 0.3 to 0.5 lb per 1000 lb of flue gas corrected to 50 percent excess air. The stack gases will be clear or nearly so and well within requirements.

BURNING WIRE INSULATION

Large tonnages of scrap wire are salvaged and returned annually to the copper refineries. The simplest way to remove insulation from quantity lots of copper wire is by burning. However, open burning of the wire insulation produces fumes and black smoke, a practice that is intolerable in or near urban areas and forbidden in some states.

The organic insulations on wire include rubber, polyvinyl chloride (PVC), neoprene, polyethylene and other plastics, as well as paper, fabric, and lacquers. Compounded with the plastics are various amounts of metallo-organic pigments, plasticizers, fire retardants, and fillers. Depending on the method of burning, the emissions to the atmosphere may include carbonaceous smoke, chloride fumes, and ash dusts among other pollutants [1]. However, the insulation can be baked to a brittle substance or char by heating without burning and then removed by flailing or other mechanical action. In this way there is minimum oxidation of the copper. However, the equipment and controls required are more costly and extensive than for direct burning, and therefore baking is economical only on the larger salvage operations.

Burning the insulation in batch or conveyorized furnaces is done best in layers not over 8 to 10 in. deep and supported on open gratings. The bales of wire must be opened so that air can flow freely through the wire to support combustion. The loose wire weighs about 14 to 15 lb/ft^3. A satisfactory layer for burning purposes weighs about 10 lb/ft^2. Lead sheathing should be stripped off beforehand. Heavy cable with steel armour is best cut into lengths of about 1 to 3 ft and mixed with lighter wire scrap. Batch charges may be ignited with torches or by lighting fuel oil sprinkled on part of the wire. Ignition spreads through and across the layer without further aid. In continuous-conveyor furnaces an oil or gas burner and a refractory arch over the incoming wire induces ignition by heat radiation from above.

Air to sustain combustion and cool the grate is supplied by natural draft below as well as over the top of the wire. After the insulation finishes burning the upward flow of air cools the wire.

Yields

The amount of copper in a ton of insulated wire varies widely but can be determined for each type of wire by weighing hand-stripped samples. Approximately 4 percent of the copper is lost in the fume and dust. In a test of several thousand pounds the average yield of No. 2 grade copper wire was 53.5 percent. The copper-bearing ash, largely from minerals in the insulation, was 10.8 percent. The weight loss to the flue gases by combustion was 35.7 percent. The copper content of the ash alone is sufficient to warrant recovery by copper refineries.

Air-Pollution Control

During the first 30 min of insulation burning dense black smoke is produced, followed by a period of gradual clearing to white and light-colored fumes. The black smoke can be eliminated by passing the combustion gases through an afterburner to attain a 1500°F exhaust temperature. An afterburner similar to the one used in burning automobile bodies would be suitable. This two-stage burning process with its low-temperature stage for destroying the insulation and its high-temperature afterburner stage for burning the hydrocarbon smoke protects the copper from severe oxidation and embrittlement. After the hydrocarbons and soot have been burned out of the gases, chlorides and mineral dust, which have been disguised in the black smoke, still remain in the off-white fume. These substances are toxic and must be removed to prevent air pollution. They result from chemical reaction between the hydro-chloric acid and the metallic compounds in the insulation. In our tests the dust and fume content averaged 240 lb per ton of original wire. The dust, consisting mainly of metallic chlorides, was 85 percent water soluble.

The amount of fume is high because the principal insulation material, polyvinyl chloride ($CH_2:CHCl$), is 56.8 percent chlorine, or 58.5 percent hydrochloric acid (HCl). Polyvinylidene chloride is 73.2 percent chlorine, and neoprene is 40.1 percent chlorine. Polyethylene is a hydrocarbon without chlorine. Teflon insulation contains the toxic gas fluorine, which reacts much like chlorine when burned.

Exhaust gases from the afterburner are cleaned best by a gas scrubber of acid-proof construction and at least 90 percent collection efficiency. The chlorides dissolve in the scrubber water. The hot gases from the afterburner are cooled and saturated with water vapor by spraying them with an excess of water before they enter the scrubber. Several types of scrubbers are available from several manufacturers. They all provide for turbulent mixing of the gases with water to produce intimate contact of the contaminants with the water. The water supplied equals the water evaporated plus the effluent water containing the chlorides and dust.

The effluent water from the scrubber is acidic and should be neutral-ized with lime or other low-cost alkali. During the neutralization a dark precipitate of metal hydroxides forms, leaving a clear solution of calcium chloride. Magnesium, sodium, and potassium chlorides are also present, depending on the neutralizing agents used. The precipitate may have salvage value; the clear liquor remaining is disposable in accordance with local regulations.

During our tests the gas scrubber operated with a pressure drop of of 42.7 in. water gage (1.54 lb/in.²). It had a dust-collection efficiency of 94.5 percent. The water became black and acidic—hence the need for an acid-proof scrubber. The hydroxides precipitated totalled 151 lb per ton of insulated wire. Insoluble solids amounted to 32 lb per ton of wire; 70 lb of lime (CaO) were required per ton of wire; 138 lb of calcium chloride were left in solution in the clarified scrubber water. The cost of wire burning is 1 to 2 cents/lb, or $20 to $40 a ton of insulated wire.

FIRE CLEANING STEEL CONTAINERS

Fifty-five gallon steel drums with removable tops are often salvageable by cleaning and repainting. When they have been used for something like oil, tar, or adhesive the most economical process for cleaning is burning. When it is done in closed batch or conveyor furnaces with temperature and combustion control, there is no smoke.

Figure 8–4 illustrates the basic principles of a continuous drum burner. The open-end furnace is refractory lined, with heat supplied by gas burners over the drums. The radiant heat from the burner and brickwork ignites the drum contents and the exterior paint. Air for combustion and for cooling the burned drums enters through the open ends of the furnace. Drums (1) and covers (2) are placed on conveyor (3), which moves continuously through the furnace at variable speed. Gas or oil burners (4) provide heat for ignition by radiation and for afterburning in the exhaust duct. Exhaust fan (6) draws air into the furnace at its open ends (5). Air, bled into the exhaust duct (7), maintains safe temperatures in the exhaust fan. The smoky products of combustion of the paint and drum residuals leave the primary chamber through ports (8) and burn out in the refractory passages (9), which are maintained at temperatures of up to 1500°F to insure smokelessness.

The capacities, maximum gas-burner inputs, and exhauster sizes for one make of drum burners are given below.*

Drums (per hr)	Maximum Gas Input (million Btu/hr)	Exhaust Blower (ft³/min)
120	3	6400
60–70	2	4000
40–50	1.8	3000

Two men can operate a furnace cleaning up to 200 drums an hour. A unit with a capacity of 120 drums per hour is said to cost about $25,000.

DEBONDING BRAKE SHOES

Metal brake shoes of passenger cars and trucks are salvageable. Freeing them of the linings and cement is readily accomplished by controlled burning in a debonding furnace. Uncontrolled burning of linings

* Courtesy of I. J. White Corporation, Flushing, New York.

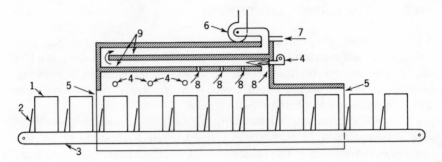

Figure 8–4 Schematic longitudinal section of drum burner. 1. Drum. 2. Drum cover. 3. Conveyor. 4. Gas or oil burners. 5. Open ends. 6. Blower. 7. Air port. 8. Gas ports. 9. Refractory passages.

and cement in the open throws off dense smoke that fills the shop and surrounding area; controlled burning in the proper furnace results in a clear exhaust gas. As shown in Figure 8–5, worn shoes (1) are attached to fixtures on conveyor (2), which moves into furnace heated by gas burners (3). The organic bonding ignites and burns. The smoky gases pass through ports (4) to burn out in the refractory-lined passages (5). The 1500°F combustion gases, mixed with cooling air from inlet (6), are exhausted by blower (7). The debonded and degreased shoes are removed at (8). Debonder furnaces with a capacity of 500 shoes per hour cost $5000 to $6000.

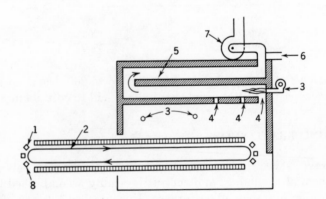

Figure 8–5 Brake-shoe-debonding furnace. 1. Worn shoes. 2. Conveyor. 3. Gas or oil burners. 4. Gas ports. 5. Refractory passages. 6. Air inlet. 7. Blower.

SWEATING FURNACES

It is possible to melt lead sheathing from insulated wire cables and to burn the insulation simultaneously in the same furnace. The cable is cut into lengths, stacked on pallets, and then loaded into the furnace by fork-lift vehicles. The furnace may be preheated before the first load is charged. Since lead melts at 620°F and the burning wire insulation provides heat, successive charges may require little or no additional gas or oil heat. A sloping refractory hearth will allow continuous tapping of the molten lead and thus reduce oxidation losses.

Figure 8–6 illustrates a sweating furnace that is suitable for aluminum melting. Locating the gas offtake near the hearth rather than at the top of the furnace allows the air leaking through cracks around the charging door to flow across the hearth and into the offtake with minimum cooling effect; also, it allows the hot gases from the burners to sweep the furnace and to heat the charge before it reaches the flame port. If the gas offtake were at the top of the furnace, the hottest gases would be near the exit, and the coldest gases around the charge.

The combustion of the smoke from the insulated wire is completed in an afterburner chamber following the sweating furnace, before the exhaust gases enter the stack. A gas or oil burner in the afterburner chamber provides any additional heat required.

This same type of furnace can be used to melt aluminum housings off transmissions or to melt scrap aluminum castings and rolled shapes. The aluminum melts when it reaches a dull red heat, at 1220°F, and drains off the iron and steel parts with which it has been associated. This saves

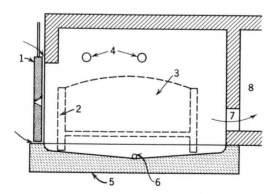

Figure 8–6 Sweating furnace. 1. Charging door. 2. Steel pallet. 3. Charge. 4. Gas or oil burners. 5. Hearth. 6. Tap hole. 7. Exhaust port. 8. Afterburner.

the cost of manual disassembly. Any smoke from grease and oil burning off the metal in the primary chamber is also consumed in the afterburner.

SUMMARY

The general principles of incineration in metal recovery apply to several salvage operations that provide an economical means of cleaning metal by combustion of organic material. The method could also apply to recovery of silver from photographic paper and film, and recovery of lead from batteries. To reduce air pollution from such operations afterburners, exhaust-gas scrubbers, and other controls may be necessary. It is wise to consult with specialists in the design of furnaces and become acquainted with modern air-pollution-control and governmental requirements before deciding on a new installation.

REFERENCES

[1] Kaiser, E. R., and J. Tolciss, "Control of Air Pollution from the Burning of Insulated Copper Wire," *J. APCA,* **13** (1) 5–11 (January 1963).
[2] Kaiser, E. R., and J. Tolciss, "Smokeless Burning of Automobile Bodies," *J. APCA,* **12** (2), 64–73 (February 1962).

9

INCINERATION OF RADIOACTIVELY CONTAMINATED COMBUSTIBLE WASTES

Richard C. Corey

Wherever radioactive materials are handled or used, radioactively contaminated solid wastes are produced. Such wastes are generated by installations engaged in research, development, and production for the nation's atomic energy program and by so-called off-site facilities, such as institutional, commercial, and industrial users of radioisotopes. The scope of this discussion is confined to off-site combustible wastes that must be disposed of safely and economically.

Unless unlimited burial or storage facilities are available to off-site facilities or it is convenient to transport the wastes to storage depots, incineration of combustible wastes may be employed. There is still a disposal problem, but its magnitude is lessened by the fact that the volume of the original waste is greatly reduced.

Solid off-site wastes generally are a heterogeneous mixture of contaminated combustible and noncombustible materials. Typical com-

Research Director, Pittsburgh Coal Research Center, Bureau of Mines, U.S. Department of the Interior, Pittsburgh, Pennsylvania.

bustible wastes are paper and paper products, plastics, wood, apparel, garbage, plants, animal cadavers, viscera, cage cleanings, and floor sweepings; noncombustible wastes may include a variety of glass, ceramics, and metals.

The volume and weight of off-site combustible wastes varies with the size of the facility and the nature of the work. The bulk density of the waste may vary from about 3 to 10 lb/ft^3, and the moisture content may be as much as 75 percent.

Because of these wide variations in the characteristics of the wastes, either within a given facility or among facilities, the incinerator must be carefully selected to achieve two results: (a) maximum combustion efficiency, with only negligible quantities of smoke, tars, malodorous gases, and/or vapors discharged to the atmosphere; (b) maximum retention of solids within the system, with a minimum amount of particulate matter entrained in the flue gases.

Tantamount to the problem of selecting a suitable incinerator to achieve such conditions are problems of personnel safety and contamination of the public domain by radioactive isotopes present in the original waste and carried over into the solid and gaseous products of incineration.

All personnel engaged in preparing radioactive wastes for incineration, operating the incinerator, or handling residues from the incinerator are potentially subject to the hazards associated with contact with radioisotopes. They must conform to recommended standards for *occupational-exposure levels*, including badging and monitoring; whereas personnel exposed to radioactivity from stack effluents must conform to standards for *nonoccupational levels*.

If metal receptacles are provided for collecting contaminated combustibles wastes exclusive of animal and pathological remains, cage cleanings, and very bulky materials, they should be provided with disposable inserts made of combustible material, such as cardboard. Clothing, bedding, and other bulky wastes should be collected in large bags provided with a closure, such as drawstrings. Animal and pathological wastes should be collected in relatively rigid combustible containers provided with close-fitted covers.

Transportation of the packaged wastes to the incinerator should be done in a manner that will avoid both spillage of waste and contact of personnel with the packages. Ample space should be provided at the incinerator site to store charges that have a relatively short half life. Putrescible charges should either be refrigerated if storage is necessary or incinerated as soon as possible.

Several studies, which are discussed later, have been made of the hazards associated with handling ash residues. In general the ash should never be disturbed in a manner that might result in its becoming airborne outside of the incinerator. The best practice is to design the incinerator so that the residue drops directly into a container that would be used for ultimate disposal by land burial, sanitary fill, transfer to waste-disposal depot, or other acceptable means of disposal. When such practice is not feasible personnel should wear approved respirators. It is generally good practice to utilize garments, such as coveralls and gloves, specifically identified for this purpose and isolated from uncontaminated clothing. The ash should be wetted to prevent its becoming airborne or removed by a vacuum-cleaning device with receptacles that can be disposed of safely.

Contamination of the public domain occurs when precautions are not taken to control the emission of radioisotopes. Since incineration does not change the radioactivity of the isotopes, the original amount will be distributed among the ash residue, the effluent from the incinerator, and adherent deposits inside the incinerator. Certain radioisotopes will appear in the flue gas; for example, ^{14}C will be discharged as radioactive carbon dioxide, $^{14}CO_2$. Others—such as the radioisotopes of calcium, ^{45}Ca and ^{47}Ca, and phosphorus, ^{32}P—will be largely retained in the ash residue unless they escape as fly ash from the incinerator stack. Radio-iodine, ^{131}I, may appear in the ash and the flue gas or it may be "plated out" on the internal parts of the incinerator.

The National Council on Radiation Protection and Measurements is preparing a handbook dealing with the biological considerations for the incineration of radioactive wastes and will establish guidelines for the maximum permissible concentration (MPC) of radioisotopes in connection with the incineration of off-site wastes. Generally, radioactive effluents released by off-site incinerators should not exceed one-tenth the exposure levels for nonoccupational exposure, as given in *Handbook 69, Maximum Permissible Body Burdens and Maximum Permissible Concentrations of Radionuclides in Air and Water for Occupational Exposure*, National Bureau of Standards.

INCINERATORS FOR CONTAMINATED OFF-SITE WASTES

Incinerators designed specifically for off-site wastes are not available commercially at the present time, although some experimental work has been done to adapt conventional incinerators to off-site requirements. Figure 9–1 shows some of the basic types of conventional incinerator. Also, a considerable amount of research has been done to establish some basic guidelines for equipment design or selection.

A brief review of research related to the incineration of contaminated

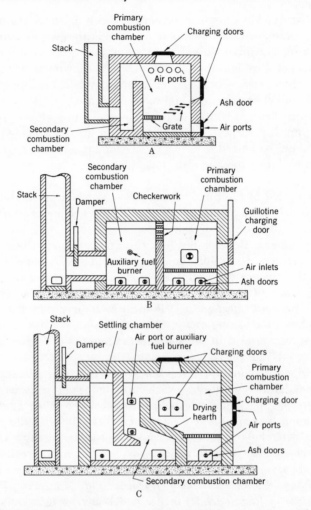

Figure 9–1 Basic design of conventional incinerators for commercial, industrial, and institutional wastes.

wastes from off-site operations will provide background information, which is necessary to understand the present state of the technology. The United States Bureau of Mines was the first to investigate incineration systematically, with the objective of furnishing basic information on the design of incinerators for contaminated wastes. The results were published in a series of papers. The first reviewed some of the basic principles of combustion, as related to incineration in general [1]. Two subsequent papers dealt with experimental studies of a variation of the underfeed combustion principle [2, 3]. Because of certain desirable characteristics of free vortex flow (i.e., flow with constant angular momentum) studies were made in a cylindrical combustion chamber to which air was admitted through tangential ports above the charge. Conclusions drawn from these studies are as follows:

1. Efficient combustion of low-ash, high-volatile wastes can be achieved with tangential overfire air by controlling the Reynolds number of the air in the ports. Undergrate air may be used if increased burning rate per square foot of grate area is desired, but undergrate air must be restricted to a relatively small fraction of the total air.

2. Efficient operation can be achieved with comparatively low excess air, which is an important factor in designing a gas-cleaning system.

3. Because of the inherent characteristics of this type of combustion system, the concentration of solids in the effluent gas will be comparatively low.

The Air Cleaning Laboratory of the Harvard School of Public Health cooperated with the Bureau of Mines during the later phases of the Bureau's work and subsequently made additional studies of cylindrical incinerators based on the design principles employed by the Bureau of Mines. The studies also included a domestic incinerator that was commercially available [6]. In its original form this domestic incinerator had a cylindrical combustion chamber constructed of steel and surrounded by a concentric steel shell with a fiber-glass lining in the annular space. A gas burner was located above the grate, and the air for combustion was a combination of underfire and overfire air with no specific entry pattern. The Air Cleaning Laboratory added a refractory lining to the combustion chamber, a predrying chamber to the top of the unit, and a slag-wool filter unit to the stack.

Burning rates averaged 25 lb/hr for sawdust and 15 lb/hr for a mixture of sawdust and shredded cabbage containing about 40 percent moisture. The particulate matter in the effluent gas averaged 0.05 gr/ft^3 and contained 20 to 40 percent of tarry products.

Although the results of these tests were encouraging, it was recognized that a prototype model of an incinerator, combining the best features of the commercial unit and a Bureau of Mines design was needed. Accordingly, the Air Cleaning Laboratory designed the unit shown in Figure 9–2. It consists essentially of a 55-gal steel drum lined with 2 in. of firebrick, providing a combustion chamber 29 in. deep and 18 in. in diameter, and an ashpit 6 in. deep. There is a single tangential overfire-air inlet with an adjustable opening and a charging door located on the side,

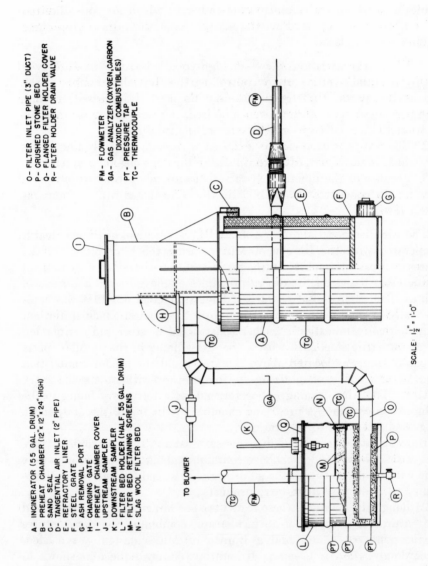

A- INCINERATOR (55 GAL DRUM)
B- PREHEAT CHAMBER (12" x 12" x 24" HIGH)
C- SAND SEAL
D- TANGENTIAL AIR INLET (2" PIPE)
E- REFRACTORY LINER
F- STEEL GRATE
G- ASH REMOVAL PORT
H- CHARGING GATE
I- PREHEAT CHAMBER COVER
J- UPSTREAM SAMPLER
K- DOWNSTREAM SAMPLER
L- FILTER BED HOLDER (HALF, 55 GAL. DRUM)
M- FILTER BED RETAINING SCREENS
N- SLAG WOOL FILTER BED

O- FILTER INLET PIPE (3" DUCT)
P- CRUSHED STONE BED
Q- FLANGED FILTER HOLDER COVER
R- FILTER HOLDER DRAIN VALVE

FM- FLOWMETER
GA- GAS ANALYZER (OXYGEN,CARBON
       DIOXIDE, COMBUSTIBLES)
PT- PRESSURE TAP
TC- THERMOCOUPLE

TO BLOWER

SCALE: $1\frac{1}{2}" = 1'-0"$

Figure 9–2 Institutional-type incinerator for disposal of low-level activity wastes.

244

6 in. above the grate level. A cylindrical, brick-lined chamber, 20 in. high and 11 in. inside diameter, and mounted directly above the combustion chamber, serves as an afterburner. A port at the base of the afterburner admits secondary, or dilution, air tangentially to the wall.

Later the Air Cleaning Laboratory designed a 50-lb/hr incinerator for installation in the United States Army Chemical Center, Nuclear Defense Laboratory, Edgewood, Maryland [4, 5]. This incinerator is a scaled-up version of the one shown in Figure 9–2. The gas-cleaning system is composed of three basic units:

1. A medium-velocity inertial cyclone-type separator.
2. An electrostatic precipitator.
3. A mineral-fiber filtration unit.

Performance tests were made on the system with uncontaminated waste to determine if the components met design specifications [10]. Subsequently, extensive performance tests were made with sawdust and carcasses on a modified system consisting of (a) a primary and secondary combustion chamber, constructed of mild steel and lined with refractory; (b) a medium-velocity, involute-cyclone separator; and (c) an electrostatic precipitator. Overfire air was admitted through tangential ports in both the primary and secondary combustion chambers. Incineration of charges containing up to 200 microcuries of ^{89}Sr per kilogram was accomplished without exceeding the maximum-permissible-concentration limits [9].

The New York Operations Office (NYO) of the Atomic Energy Commission has reported on the use of conventional, industrial-type incinerators by three of its contractors to dispose of uranium-contaminated paper, rags, gloves, towels, boxes, etc. [4].

One of the units was rated at 210 lb/hr; it had an auxiliary settling chamber for fly ash and a short stack. No gas-cleaning equipment was provided. Radioactivity was monitored with a sampler in the stack and four samplers placed at intercardinal points 3 ft above and 8 ft away from the stack. The cost of the installation was estimated at $3000. Another unit was rated at 200 lb/hr and had a 1000-cfm centrifugal dust collector and a 20-ft stack. The effluent was monitored with a sampler a few feet from the stack. The cost of the installation was estimated at $3500. The third unit was rated at 600 lb/hr and had a 35-ft stack. No gas-cleaning equipment was provided. Monitoring was done with samplers in the stack, at distances 150 to 500 ft from the stack.

None of the units exceeded the maximum permissible concentration for either 40-hr occupational exposure or 168-hr nonoccupational exposure during extensive tests. Pending more careful study of the data on hand and additional tests it was tentatively concluded that incineration of combustible wastes from reactor-fuel fabrication can be done at comparatively low cost, with high retention of radioactivity in the ash residue and insignificant air contamination. It was hoped ultimately to be able to recommend commercial incinerators as a primary method for disposing of combustible wastes from fuel-fabrication facilities.

Several years ago the University of California at Los Angeles experimented with a commercial, gas-fired, domestic incinerator for the disposal of medical and biological wastes [13]. Over a period of 286 burn days 3400 lb of assorted wastes with an average activity of 95 microcuries were burned. The flue of the incinerator was connected to a stack approximately 25 ft high. No special means were employed to clean the stack gases, but precautions were taken to protect the operator from airborne ash while the ash receptacle below the grate was being cleaned. The stack-gas beta radioactivity was normally less than 7×10^{-11} microcuries/cm^3, and no activity above normal background levels was detected in areas surrounding the incinerator.

Johns Hopkins University investigated the performance of a conventional incinerator in its School of Hygiene and Public Health, while burning radioactive wastes [11]. The incinerator design was approximately that shown in Figure 9–1C and had a rated capacity of 150 to 200 lb/hr. The stack was 143 ft high. Known quantities of radioisotopes were charged to the incinerator in amounts ranging from 200 to 2000 microcuries for each experiment. The charge was added as chick embryos or on cotton and sawdust wetted with the isotope solution. Measurements of radioactivity were made on the stack gases, the atmosphere near the stack, the stack walls, and the ash residue.

With ^{32}P and ^{89}Sr about 90 percent of the isotope remained in the ash, where it might produce a hazard during ash removal and disposal. With ^{131}I about 80 percent entered the stack gases; however, to reach undesirable atmospheric concentrations, the ^{131}I loading would greatly exceed the amount permitted for safety in handling.

The Bureau of Industrial Hygiene, Detroit, Michigan, designed a small stainless-steel, rotating-drum incinerator for wastes containing microcurie amounts of ^{14}C [11]. The drum was 18 in. in diameter and 24 in. long; it was fired at one end with natural gas. The products of combustion issued from the other end of the drum and entered a wet-type fly-ash collector $16\frac{1}{2}$ ft wide, 26 ft long, and $27\frac{1}{2}$ in. high that also served a large incinerator used for uncontaminated wastes.

Five-pound bags of small animal carcasses were charged to the drum, which was preheated to about 1800°F. Incineration was completed in about 25 min. The residual ash was found to contain less than 0.008 percent of the activity as charged. Gas samples indicated a concentration of 0.084 microcurie/cm³ at an incineration rate of 1.35 curies/min. The authors concluded that this method is probably safe for substantially higher rates of incineration.

The Harvard School of Public Health studied dust exposures during the removal of ash from six conventional institutional or municipal incinerators for disposing of contaminated wastes [12]. The conclusions were as follows:

1. Dust exposures to the operator during ash removal are not excessive with the average institutional-type incinerator, because the stack draft through the open furnace doors draws most of the dust produced by the operation back into the incinerator. Installations with small stacks (or when reverse flow is created by wind action) produce higher and more persistent dust concentrations. Several operators commented that incinerator operation was poorer during rainy weather. The effect of outside weather conditions on draft and ventilation of the cleaning operation is apparent.

2. Incinerators of the type tested, with water quenching of ash before discharge to the ash truck, produce only minor dust exposures. Where feasible, wetting the ash is advisable.

3. Airborne dust in incinerator rooms is of the same general composition as the ash. It contains a high percentage of small particles (50 percent less than 0.5 micron); larger particles are predominantly agglomerates of the smaller particles. Enough of the larger particles are airborne to cause a large standard geometric deviation of the size-distribution curve.

4. Exposure period for operators removing ash is a function of the degree of cleaning accomplished as well as the quantity of ash removed.

5. Stack draft and room ventilation appear to be the most important factors in limiting the concentration of airborne dust in the incinerator room.

6. Much dust exposure incident to ash removal can be eliminated by careful placement of disposal cans to take advantage of the effect of stack draft and by care in depositing ashes from the shovel into containers. Obviously the furnace door and grate opening should be kept open throughout all cleaning in the incinerator room.

WASTE DISPOSAL IN A BOILER PLANT

An efficient solid-fuel-boiler plant, furnishing heat and power to an institution or other user or radioisotopes, offers possible means for contaminated waste disposal. Waste charged to such a plant would be considerably dilute since the ratio of boiler fuel to waste would be quiet high as long as the waste was charged to the furnace in prescribed increments and not as a large, single batch.

If a boiler furnace is to be used for contaminated-waste disposal, the following factors should be considered:

1. Safe handling of the waste in the plant.
2. Safe means for charging the packaged wastes to the furnace.
3. Scheduling the charging to avoid feeding large charges in a short time.
4. Monitoring the stack gases and areas where residues are collected or tend to accumulate until it has been demonstrated that hazards do not exist.

A comparison was made of the performance of a crematory-type incinerator and a combination steam boiler-incinerator for the combustion of contaminated laboratory and hospital wastes, such as animals and cage litter [7]. Because of excessive stack emissions of smoke, fly ash, and malodorous gases and vapors the crematory unit proved unsatisfactory. The mechanized steam boiler-incinerator, on the other hand, provided a sanitary method for handling and burning such wastes in an efficient and rapid manner. A cyclone dust collector proved superfluous for use with gas or oil fuels, and it could be eliminated without decreasing the efficiency of an electrostatic precipitator, which served as a final cleaning stage.

The steam boiler-incinerator is shown schematically in Figure 9–3. It consists of a heavy-oil-fired boiler modified to burn packaged solid materials on a specially constructed hearth. An automatic system, shown in Figure 9–4, was employed to charge packages into the combustion chamber.

Radioactive tests were made with known amounts of tritium, ^{14}C, and ^{35}S, which were added to cage litter and animals to determine the release rates of these substances from the charge and their distribution among flue gases, residue, furnace linings, and duct surfaces.

Because of the efficiency of incineration and the enormous isotope dilution that occurs, the authors concluded that there is no likelihood of exceeding the maximum permissible release rates to the atmosphere of these commonly used radionuclides. However, radioactive material concentrated in the solid residue, external exposures of plant personnel, and permissible levels for disposal in public dumps require careful consideration.

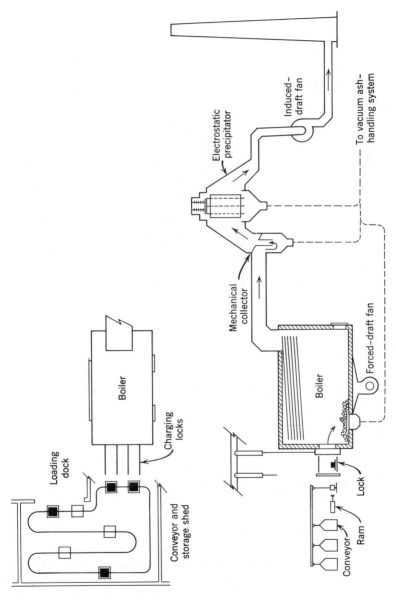

Figure 9-3 Layout of incinerator-boiler system at Harvard University Medical Center [7].

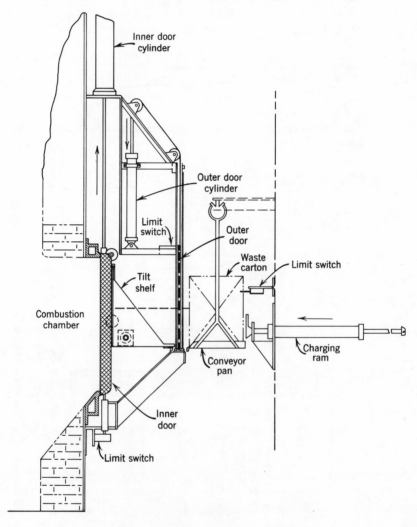

Figure 9–4   Waste-charging door [7].

CLEANING STACK EFFLUENTS

The need for removing particulate matter from the products of combustion depends mainly on the amount of radioactivity that is to be charged to the incinerator. The quantity of gas and the particulate loading determine the type and size of gas-cleaning equipment that is necessary to keep the activity of the effluent within permissible limits.

When high dust loadings with a high level of activity are anticipated a cleaning system is essential. In any new design it would be wise to include space and provisions for connections for a cleaning system if it is found later that one is needed.

Wet-collection methods require both wet and dry disposal facilities. Consequently, it is preferable to use dry-type cleaning devices, such as cyclones, filters, electrostatic precipitators, or various combinations of these types of devices. Dry filters have temperature limitations; the temperature of the gas must not be below its dewpoint if condensate would damage the filter material or exceed the limit of the filter material toward melting, charring, or burning. Cyclones and electrostatic precipitators operate up to about 1000°F, but electrostatic precipitation may be difficult to justify for a small waste-disposal facility because its capital cost is comparatively high.

The particulate loading can be reduced 50 to 75 percent by cyclones and 90 percent or better by a cyclone-precipitator combination. A roughing filter followed by an absolute filter removes up to 99.9 percent of the particulate matter. Such filters can be burned in the incinerator and the ash can be removed with the other residues.

SUMMARY AND CONCLUSIONS

The principal problems associated with disposal of low-level contaminated wastes generated by institutional, commercial, and industrial users of radioisotopes are the following:

1. Selection of a suitable incinerator with regard to type and size.
2. Protection of personnel handling the raw wastes and the solid residues from the incinerator.
3. Prevention of contamination of the public domain by excessive radioactivity in the gaseous and particulate emissions from the incinerator stack.

If incineration is selected as the best method for waste disposal, the following choices are available:

1. Selection of a commercially available incinerator suited to the kind and amount of waste that is to be handled.
2. Modification of such an incinerator to incorporate features necessary to convenience, safety, and adaptability to changing conditions.
3. Design and construction of an incinerator for the specific job.

Factors dictating the size and type of an incinerator, and the decision whether or not auxiliary firing is needed are the following:

1. The estimated amount of waste to be burned in a given time, which fixes the dimensions of the combustion chamber and the grate and/or hearth.

2. Whether the waste is relatively dry (up to about 25 percent moisture) or wet and slow burning, which determines whether auxiliary firing is needed, and if so, the best arrangement of the burners and the optimum hourly fuel rate.

3. The bulk density of the waste, which determines the method of charging to the incinerator, the dimensions of the charging aperture, and the kind of grate and/or hearth in the combustion chamber.

Whether or not gas-cleaning equipment is needed depends on the kind and the concentration of the radioisotopes in the waste. Generally it is best to provide for gas cleaning in the system and to install it later if careful monitoring of the stack gases and the surroundings indicate potential or real hazards. Otherwise it may be found that the gas-cleaning system is either unnecessary or inadequate with respect to size and kind. Under no circumstances should a gas-cleaning system permit discharge of contaminated effluent to municipal waste-water or sewage facilities.

Some of the problems associated with discharge of radioisotopes to the atmosphere can be solved or mitigated by isotopic dilution of the waste, that is, by burning a small amount of contaminated waste with a comparatively large amount of uncontaminated waste. Disposal of the waste in a boiler plant would effect isotopic dilution because of the high ratio of boiler fuel to waste. Diluting the waste charged to any incinerator with a large amount of sawdust, paper, etc. would have the same effect.

At the present time the state of the technology is not sufficiently advanced for a packaged system to be installed. Each situation must be considered as different from any other one; this means that each step from collection of the waste, through incineration, to disposal of the final residue must be examined in detail and made compatible with all the other steps.

REFERENCES

[1] Corey, R. C., "Fundamental Considerations of Design and Operation of Incinerators in Relation to Atmospheric Contamination," AEC Waste Processing Committee Meeting, Los Alamos, New Mexico, October 1950; TID–460, pp. 58–69.

[2] Corey, R. C., "Process Variables Study of Incineration Using Tangential Overfire Air," Air Cleaning Seminar, Ames Laboratory, September 15–17, 1952; WASH–149, March 1954, pp. 154–169.

[3] Corey, R. C., L. A. Spano, C. H. Schwartz, and H. Perry, "Experimental Study of Effects of Tangential Overfire Air on the Incineration of Combustible Wastes," *Air Repair* (JAPCA), **3**, 1–8 (1953).

[4] Dennis, R., and L. Silverman, "Radioactive Waste Incinerator Design and Operational Experience—A Review," AEC Div. Tech. Inf. TID–7627, Book 2, pp. 416–460.

[5] Dennis, R., W. Kyritsis, and E. W. Bloore, "An Evaluation Program for Radioactive Waste Incineration," U.S. Army Nuclear Defense Laboratory, NDL–TM–8, March 1963.

[6] Dennis, R., F. L. Muller, E. Kristal, and L. Silverman, "Special Incineration. Studies—Institutional Design," Sixth AEC Air Cleaning Conf., Idaho Falls, July 7–9, 1959; TID–7593, Washington, D.C., October 1960, pp. 344–364.

[7] First, M. W., P. Zilles, and J. Walkley, "Disposal of Radioactive Waste in Commercial Incinerators," Ninth AEC Air Cleaning Conf., Boston, September 13–16, 1966, pp. 570–585.

[8] Kruse, C. W., P. V. Freese, A. Machis, and V. C. Behn, "Behavior of Institutional Incinerators When Used To Burn Radioactive Wastes," Johns Hopkins University, NYO–4517, November 1952.

[9] Lachapelle, D. G., "An Engineering Evaluation of a Radioactive-Waste Incinerator," Ninth AEC Air Cleaning Conf., Boston, September 13–16, 1966, pp 509–569.

[10] Lachapelle, D. G., J. L. Tarbox, and D. L. Goff, "Evaluation Program for Radioactive Waste Incineration," U.S. Army Nuclear Defense Laboratory, NDL–TM–24, October 1965.

[11] Lavetter, Victor E., "An Incinerator for Wastes Containing Microcurie Amounts of Carbon-14," *Ind. Hygiene J.*, 485–488 (December 1961).

[12] Megonnell, William H., John H. Ludwig, and Leslie Silverman, "Dust Exposures During Ash Removal from Incinerators," *AMA Arch. Ind. Health*, **15**, 215–222 (March 1957).

[13] Silverman, Louis B., and R. K. Dickey, "Reduction of Combustible, Low-Level Contaminated Wastes by Incineration," Atomic Energy Project Contract AT–04–1GEN–12, University of California School of Medicine, Rept. UCLA 368, May 15, 1956.

10

TESTING INCINERATOR PERFORMANCE

Ralph E. George and John E. Williamson

In any progressive incinerator program it is necessary to know the type and quantity of contaminants emitted into the atmosphere from incinerators of different designs. Such data are important to the designer, to the manufacturer, and to air-pollution-control officials.

Many government agencies as well as a large segment of industry have established maximum limits of emission for specific contaminants. As air pollution becomes an increasing danger to public health, promulgation of new air-pollution-control districts and increasingly stringent regulations is inevitable. In most cases the only way of determining compliance with most of these regulations is by source testing.

Information obtained by source testing is also invaluable in selecting appropriate control equipment and in aiding designers of control equipment to minimize emission of air contaminants. It is the only reliable method of determining the efficiency of air-pollution-control equipment.

Regardless of the eventual use of the information, the immediate objective of testing an air-pollution source is to obtain reliable data on the

Senior Air-Pollution Analyst and Senior Engineer, respectively, Los Angeles County Air-Pollution-Control District, Los Angeles, California.

composition of its effluent and its rate of emission to the atmosphere. The following requirements are basic to obtaining this information from any source test:

1. The gas stream being sampled from an incinerator should represent a known portion of the emission from the source.

2. The sample of the emission collected for analysis must be representative of the gas stream being sampled.

3. The volume of the gas sample withdrawn for analysis must be measured in order to calculate the concentration of the constituents in the gas stream.

4. The gas flow rate from the incinerator must be determined in order to calculate the emission rate for various constituents.

Means for insuring that these requirements are met, as well as other factors that must be considered in planning and conducting a source test, are considered in detail in this chapter.

The performance of a source test from the initial planning to the final report can be divided into a number of steps. The scope of each step depends on the magnitude and complexity of the test program. Relatively simple tests on familiar sources require much less time and effort than tests designed to yield information on new installations. These steps are listed below in the order in which they are normally taken.

1. Background of source test.
2. Inspection of source for physical requirements.
3. Selection of test procedures.
4. Scheduling of test.
5. Measurement of gas flow rate.
6. Collection of samples.
7. Processing and analysis of samples.
8. Calculation from field laboratory data and preparation of test report.

PLANNING A SOURCE TEST

Intelligent planning of a source test requires an understanding of the purposes and ultimate use of the desired test data and adequate information on the nature of the source to be tested. When the sources or equipment have not been previously tested a preliminary visit to the test site by the testing personnel can save much time and effort. The best location for gas flow measurement and sampling should be selected. In many cases it may be necessary to allow time to prepare test holes, test platforms, or scaffolds. Also to be considered during the inspection are accessibility;

space requirements; availability of electricity; the operating schedule of the equipment; and estimates of preliminary measurement of temperature, velocity, pressure, and moisture content of the gas stream. It is evident that the amount of information and planning needed will vary greatly, depending on test requirements and familiarity with the source to be tested.

At most source test locations sampling is performed at sites above the ground. The location of equipment at the sampling stations must present no undue hazard to personnel. Sufficient information should be obtained beforehand for suitable protection if the emissions are toxic, unusually hot, or under positive pressure; also if other hazardous conditions exist.

In conducting a source test two more requirements should be met:

1. The test should be conducted under representative operating conditions of the equipment.
2. Techniques for sampling, analyzing, and measuring flow rates should be carefully selected for accuracy.

There is no uniform method for either collecting or analyzing all types of air contaminants. More than one method may exist for determining any specific component of the emissions. Selection of methods may involve such factors as the physical and chemical properties of the effluent being tested and the condition of the stack gases. Other factors that may be involved in the selection of methods are limitations of working space, accessibility of test station, special equipment needed, time, and personnel.

The scheduling of the actual test, following the plant inspection, must allow time to provide test facilities, personnel, and equipment needed for tests of greater scope or complexity than the routine tests, as well as for development of special test methods. The date must also be coordinated with plant management personnel to insure operation of facilities under conditions specified for the test.

THE DETERMINATION OF GAS FLOW RATE

Test stations may be situated at the top or outlet of the stack, or at some other location in the stack. They may also be located in the ductwork leading to or within the control equipment. Frequently it may be necessary to provide test holes to allow access to the gas stream.

Gas-velocity measurements should be made at the locations where gas samples are taken, and the sampling sites should be situated at points where the gases have been well mixed and are as free as possible from distortion or nonuniformity of flow. Disturbances can be caused by a dilatation or a constriction in the stack, by changes in the direction of flow caused by bends or obstructions, or by inlet or exit gas streams in branch ducts. Normally the distorting effects of nonuniform flow are for all practical purposes damped out at a distance of 8 to 10 pipe diameters downstream from the disturbance. However, many cases will be found—usually at inlets to control equipment or for some outlet stacks —when measurements must be made at points closer to the disturbance.

Flow-measurement and sampling techniques must be carefully applied to minimize errors that may result from compromises necessary in the selection of test stations. Practically all source tests include the measurement of the flow rate of gases at one or more locations in the equipment being tested. The theory of measurement of velocity of flowing liquids and gases by use of Pitot tubes is covered extensively in standard engineering references. A commercial Pitot tube of the standard type is made of stainless steel for use at elevated temperatures or in corrosive atmospheres. As shown in Figure 10–1, it consists of two concentric tubes bent at a right angle, with the inner tube open at the tip.

The opening of the inner tube, when directed against the gas flow, measures the total (also called impact or stagnation) pressure of the flowing gas. The outer tube measures the static pressure of the gas through

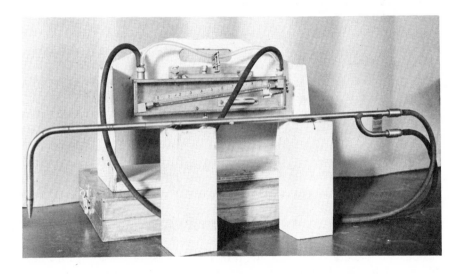

Figure 10–1 Gas-velocity-measuring equipment with an inclined manometer and standard Pitot tube.

a number of small holes drilled through the tube wall between the tip and the bend, perpendicular to the impact opening. The difference in pressure between the two tubes is equal to the velocity pressure, or velocity head, of the gas at the point of measurement. This pressure difference is measured with an inclined manometer connected to the taps of the Pitot tube, as also shown in Figure 10–1. The inclined manometer must be sensitive to pressure differences of 0.01 in. of water or less.

The basic relation between velocity head and gas velocity is

$$u = \sqrt{2gh},\qquad(10\text{–}1)$$

where u = gas velocity,
$\quad g$ = acceleration due to gravity,
$\quad h$ = height of a column of a liquid, equivalent in pressure to the velocity-head reading.

For gases of density equal to air and at 1-atm pressure (14.7 psia or 29.92 in. Hg barometric pressure) the above formula can be written as follows:

$$u = 2.90\ \sqrt{HT},\qquad(10\text{–}2)$$

where u = gas velocity in feet per second,
$\quad H$ = velocity head in inches of water,
$\quad T$ = gas temperature in degrees Rankine.

In cases where the gas stream is excessively dusty or laden with droplets of moisture that may clog the small static-pressure holes in a standard Pitot tube a type S (Stauscheibe) Pitot tube is used. This tube is also used in situations where increased wall thickness of the flue due to brick lining or insulation prevents the insertion of the standard Pitot tube through the opening provided for velocity measurement and sampling. The type S Pitot tube can pass through an opening as small as 1.25 in. in diameter. Since this type of Pitot tube produces a reading somewhat higher than the true velocity head, or pressure, a calibration correction (normally between 0.8 and 0.9) must be applied in calculating gas velocity from (10–2). For the standard Pitot tube the calibration factor is unity. A comparison of the static and impact opening of the standard and type S tubes is shown in Figure 10–2. The Pitot tube and inclined manometer constitute a null-balance system in which there is no flow of gas through these components. This is an advantage in the use of the Pitot tube compared with other flow-measurement devices when hot or corrosive gases are present. A disadvantage of the Pitot tube is the relatively low velocity head produced (e.g., a reading of only 0.01 in. of water for a velocity of

Figure 10–2 Tips of two common types of Pitot tube. A standard type (left) and a Stauscheibe type (right).

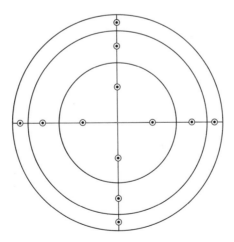

Figure 10–3 Cross section of a circular stack divided into three equal areas. The center of equal areas is indicated by ⊙.

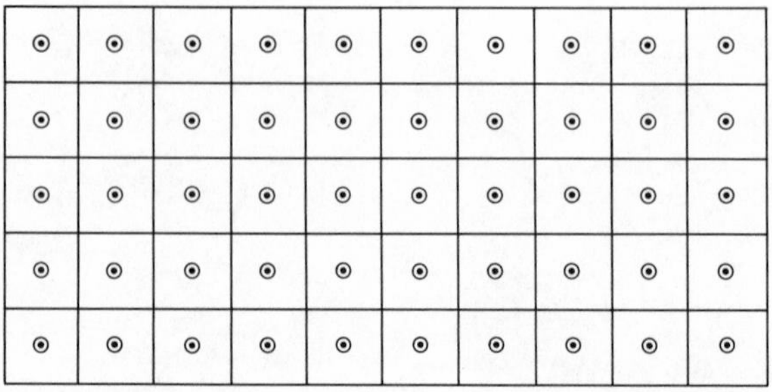

Figure 10–4 Cross section of a rectangular stack divided into 12 equal areas. The center of equal areas is indicated by ⊙.

9 ft/sec for air at 500°F). For lower gas velocities devices such as sensitive micromanometers and pressure transducers may be used if conditions at the sample station locations permit.

The Pitot tube measures velocity head only at the point in the gas stream where it is placed; therefore readings must be made at a number of points in the stack cross section so that the average gas velocity may be calculated. The number of points to be used and their location can be readily determined in accordance with commonly accepted practice.

The cross-sectional area of the stack at the test station is divided into a number of concentric equal-area zones (as shown in Figure 10–3 for circular stacks) or rectangular equal areas (as shown in Figure 10–4 for rectangular stacks). The number of areas used depends on the flow pattern and the size of the stack. For circular stacks with fairly uniform flow the usual practice is to employ the number of areas shown in Table 10–1.

Table 10–1 Suggested Number of Equal Areas for Velocity Measurement in Circular Stacks

Stack Diameter (feet)	Number of Equal Areas
1 or less	2
1–2	3
2–4	4
4–6	5
more than 6	6 or more

The traverse points for velocity-head measurements are located in the center of each area, as indicated in Figures 10–3 and 10–4. These points are at the center of each rectangular area for rectangular ducts and stacks. For circular stacks the location of each point along a stack diameter is calculated from the expression

$$P = 50 \left[1 - \left(\frac{(2j - 1)}{2a} \right)^{\frac{1}{2}} \right], \qquad (10\text{–}3)$$

where P = percent of stack diameter from inside wall to traverse point,
a = total number of areas being used,
j = number of the area for which the location is being calculated, 1, 2, 3,... from the center outward.

This formula gives half the values needed, the remaining half being the difference of each percentage from 100. In practice it is simpler to use Table 10–2, in which the percentages have already been calculated for the most frequently encountered numbers of points and areas.

When test holes are necessary in circular stacks, two holes are made along diameters at right angles to each other, large enough for the insertion of the Pitot tube, sampling probes, and thermocouples. Four holes are preferable for stacks over 6 ft in diameter to avoid use of extensions on Pitot tubes. With thin-walled stacks $1\frac{1}{2}$- to 2-in. diameter holes are

Table 10–2   Percent of Circular-Stack Diameter from Inside Wall to Traverse Point

Point Number	Number of Areas Selected				
	2	3	4	5	6
1	6.2	4.4	3.3	2.2	2.0
2	25.0	14.7	10.5	8.2	6.7
3	75.0	29.4	19.5	14.5	11.8
4	93.8	70.6	32.1	22.7	17.7
5		85.3	67.9	34.4	25.0
6		95.6	80.5	65.6	35.4
7			89.5	77.3	64.6
8			96.7	85.5	75.0
9				91.8	82.3
10				97.8	88.2
11					93.3
12					98.0

sufficient. For thick walls larger holes are required unless a type S Pitot tube is used. With rectangular ducts or stacks holes are located so that the Pitot tube can conveniently traverse the centers of the rectangular areas.

In making a velocity traverse the Pitot tube is moved along the stack diameter, reading the velocity head at each traverse point with an inclined manometer. Gas temperatures, if not uniform over the stack, can be measured at each point with a thermocouple attached to the Pitot tube. Thermometers are convenient for use under uniform conditions when gas temperatures are below 700°F.

From Figure 10–3 it is seen that for circular stacks there will be four velocity-head readings for each area when readings are taken along two diameters. Normally three complete traverses are made in order to calculate a reliable average stack-gas velocity.

Before and during the traverse the following special precautions should be taken:

1. The manometer, tubing, and connections should be tested for leaks.
2. The manometer should be carefully leveled, and the liquid column accurately zeroed and freed of bubbles. (Commercial instruments usually employ colored oil of known specific gravity with the scale calibrated to read inches of water.) Subsequent checks should be made during the course of the test.
3. The length of the Pitot tube should be adjusted so that it can be handled easily from outside the stack when held at any traverse point. The impact opening should always point directly upstream into the flowing gases. The Pitot tube should be blown out frequently if a clogging gas stream is present.
4. The sampling port should be sealed as tightly as possible when taking velocity-head readings to minimize any effect on the static pressure of the stack gases adjacent to the opening. This is especially important when the type S Pitot tube is used.

The apparent gas velocity is calculated for each traverse point by using the following equation:

$$u_t = 2.90 \sqrt{H_t T_t}, \tag{10-4}$$

where u_t = apparent gas velocity in feet per second,
H_t = velocity head in inches of water,
T_t = gas temperature in degrees Rankine.

Numerous repetitive calculations can be eliminated by using velocity tables that list gas velocities corresponding to velocity-head and temperature readings. Such a set of tables, covering a temperature range of 10 to 1600°F and a velocity-head range of 0.001 to 2.00 in. water column, has been developed by the Los Angeles County Air-Pollution-Control District. The velocities and gas temperatures are then averaged to give the average apparent stack-gas velocity and the average stack-gas temperature from traverse-point readings.

One or more corrections to the apparent velocity are usually necessary to obtain the true stack-gas velocity. These corrections are made for the following:

1. Use of type S Pitot tube.
2. Static pressure of stack gas, if appreciably different from 1 atm (29.92 in. Hg).
3. Density of stack gas, if appreciably different from that of air at the same temperature and pressure.

The average stack-gas velocity is corrected for Pitot-tube calibration, and for pressure and density differences from atmospheric by using the following relationship:

$$u_c = u_s F_P \left[\left(\frac{29.92}{P_s} \right) \left(\frac{1.00}{D} \right) \right]^{1/2}, \tag{10-5}$$

where u_c = true or corrected average stack-gas velocity in feet per second,

u_s = average stack-gas velocity during sampling period in feet per second,

F_P = calibration factor for Pitot tube,

P_s = static pressure of the stack gases in inches of mercury absolute,

D = specific gravity of the stack gases referred to air.

Excessive calculations can be avoided by application of Pitot, pressure, and density corrections to the average velocity rather than the individual point velocities.

The value of F_P is unity for a standard Pitot tube. When a type S tube is used F_P should first be calibrated by comparing it with a standard Pitot tube over the range of gas velocities that will be encountered in testing. This can be conveniently done if a variable-draft blower system is available.

The static pressure of the stack gases usually differs little from atmospheric pressure. The difference can be measured either with the static-pressure side of the standard Pitot tube or with a regular pressure tap connected to one leg of a manometer. When static pressures differ by more than a few inches of water from atmospheric a seal between the Pitot tube and the sample port should be provided when making velocity head or static pressure measurements. The value of P_s in (10–5) is the algebraic sum of this pressure difference and atmospheric pressure. The latter value is usually close to 29.9 in. Hg for most incinerator test locations.

At the time velocity measurements are made the exact specific gravity of the gases may not be known. For field calculations of an approximate nature, such as sampling rate determinations or field checks on gas flow rates, the value of D may be taken as unity. The exact value of the term $\sqrt{1.00/D}$, called the gas-density correction factor, is calculated from the Orsat gas analysis and water-vapor content. This calculation is shown later because the value of $\sqrt{1.00/D}$ for most flue and stack gases is slightly greater than unity due to the presence of water vapor. It seldom exceeds 1.05, and for pure steam this factor is only 1.27.

To determine the average gas velocity during the test it is possible to take velocity-head and temperature measurements at traverse points during the test period and to obtain a true average in this manner. However, this is neither easy nor practical when gas velocities are continually changing and the sampling rates must constantly be adjusted.

The preferable method is to record velocity-head and temperature readings at a selected reference point in the stack, usually the center. Readings at this reference point are made alternately with traverse-point readings during the velocity-traverse procedure. The gas velocity for each reference point reading is given by

$$u_r = 2.90 \sqrt{H_r T_r}, \qquad (10\text{–}6)$$

where u_r = apparent gas velocity in feet per second,

H_r = velocity head in inches of water,

T_r = gas temperature in degrees Rankine

for each reference-point reading.

The ratio, called the flue factor, of average stack-gas velocity to reference-point gas velocity is calculated as

$$F_f = \frac{u_t}{u_r}, \qquad (10\text{–}7)$$

where F_f = flue factor,

u_t = average stack gas velocity from traverse point readings in feet per second,

u_r = average reference-point gas velocity from readings taken during velocity traverses in feet per second.

When reference-point velocity-head and temperature readings are made at intervals during the sampling period the average stack-gas velocity during sampling is given by

$$u_s = u_{rs}F_f, \qquad (10\text{--}8)$$

where u_s is the average stack-gas velocity during the sampling period, in feet per second, and u_{rs} is the average reference-point gas velocity during the sampling period in feet per second.

The corrected average stack-gas velocity during the test period may be determined by using the following formula:

$$u_c = u_{rs}F_fF_P \left[\left(\frac{29.92}{P_s}\right)\left(\frac{1.00}{D}\right)\right]^{1/2}, \qquad (10\text{--}9)$$

where the symbols are as defined for (10–5), (10–7), and (10–8).

The gas flow rate during the test at stack conditions is given by

$$Q_s = 60u_cA, \qquad (10\text{--}10)$$

where Q_s is gas flow rate at stack conditions in cubic feet per minute and A is the cross-sectional area in square feet. When corrected to standard conditions (60°F and 29.92 in. Hg),

$$Q = Q_s \left(\frac{520}{T_s}\right)\left(\frac{P_s}{29.92}\right), \qquad (10\text{--}11)$$

where Q is the gas flow rate in standard cubic feet per minute at 60°F and 29.92 in. Hg and the other symbols are as already defined.

Forms used for recording the data and calculations discussed above are shown in Figures 10–5 and 10–6. The cross-sectional-area diagrams indicate dimensions and traverse points with respect to geographical direction and orientation. Traverse-point distances should be recorded.

SAMPLE COLLECTION

The collection method used to test incinerators involves sampling a portion of the effluent over a period of time with a sampling, or collection, train. This process is continuous sampling, as distinct from grab or instantaneous sampling frequently used for gaseous constituents. The train separates the particulate matter from the gaseous components of the sample and measures the volume of gas sampled. The sampling train consists of a sampling probe and nozzle, a collection device or devices for separation of the particulate matter from the gases, a flow-metering instrument, and a vacuum pump for producing and regulating the flow of the sampled gas through the components of the sampling train. The

TEST NO. ____________ PAGE ______

SAMPLING STATION ______________________________ DATE ______

GAS-VELOCITY DATA

Time	Point	Velocity Head (in. H_2O)	Temperature (°F)	Velocity (ft/sec)	Velocity Head (in. H_2O)	Temperature (°F)	Velocity (ft/sec)	Velocity Head (in. H_2O)	Temperature (°F)	Velocity (ft/sec)

A. Indicated velocity (traverse) (ft/sec) ______

B. Indicated velocity (reference point) (ft/sec) ______

C. Flue factor, A/B ______

D. Pitot correction factor ______

E. Gas-density correction factor ______

F. Gas pressure in stack (in. Hg abs.) ______

G. Gas-pressure correction factor, $\sqrt{29.9/F}$ ______

H. Corrected velocity, $A \times D \times E \times G$ (ft/sec) ______
or $B \times C \times D \times E \times G$ (ft/sec) ______

J. Area of flue (ft^2) ______

K. Average flue temperature (°F) ______

L. Flow rate, $H \times J \times 60$ (cfm) ______

M. Flow rate, $(F/29.9) \times 520 \times L/(H + 460)$ (scm) ______

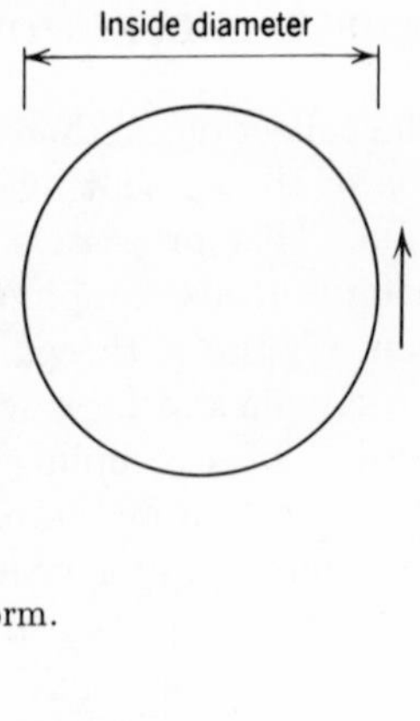

Inside diameter

Figure 10–5 Gas-velocity data form.

TEST NO. PAGE
SAMPLING STATION DATE

WATER-VAPOR AND GAS-DENSITY CALCULATIONS

PERCENT WATER VAPOR IN GASES

A. Gas pressure at meter (in. Hg abs.)
B. Vapor pressure of water at impinger temperature (in. Hg)
C. Gas volume metered (scf)
D. Water vapor metered, $C \times B/A$ (scf)
E. Water vapor condensed, vapor volume (scf)
F. Total water vapor in gas sample, $D + E$ (scf)
G. Total gas volume sampled, $C + E$ (scf)
H. Percent water vapor in gas sample, $100 \times F/G$

GAS-DENSITY CORRECTION FACTOR

Component	Volume Percent/100	Moisture × Correction $1 - H/100$	× Molecular Weight	= Weight per Mole of Stack Gas
Water	dry basis	1.0	18.0	
Carbon dioxide	dry basis		44.0	
Carbon monoxide	dry basis		28.0	
Oxygen	dry basis		32.0	
Nitrogen and inert gases	dry basis		28.2	

J. Molecular weight of stack gas
K. Density of gas referred to air $= J/28.95 =$
L. Gas-density correction factor $= \sqrt{1.00/K} =$

Figure 10–6 Water-vapor and gas-density calculation form.

collection devices may use principles of inertial separation, filtration, wet or dry impingement, electrical precipitation, liquid scrubbing, condensation, or combinations of these. The particulate matter must be collected in a state or condition appropriate for the desired use or analysis. It is evident that the collection efficiency, if not 100 percent, must otherwise be known or predictable.

The basic requirements necessary for obtaining reliable data were presented earlier. However, some of the factors enumerated require amplification at this point so that pertinent details can be recorded.

It has been stated that the sample should be representative of the total source emissions when the equipment is operating under desired conditions and must be collected over sufficient time to ensure a representative average sample.

Representative Sampling

To obtain a truly representative sample of the effluent, assuming for the moment uniform dispersion of the particulate matter in the gas stream, it is of paramount importance that the solid or liquid particles enter the sampling apparatus with the least amount of disturbance from their motion in the gas stream. This requires the sampling probe and nozzle to be pointed directly upstream into the flow of gas and the velocity of the portion of stack gas being drawn into the nozzle to be very nearly equal to the velocity of the stack gases in the vicinity of the nozzle. Sampling under these conditions is termed isokinetic sampling. If sampling is not isokinetic, segregation of particle sizes will occur—and the particle-size distribution, weight concentration, and chemical composition will not be representative. Consideration of momentum principles shows that the concentration of the heavier (and usually larger) particles is too low in the collected gas sample if too high a velocity is maintained at the sample nozzle; it is too high if the sample-nozzle velocity is lower than the local stack-gas velocity.

Deviations from isokinetic-sampling conditions may be permitted when maximum particle sizes are 5 microns (0.005 mm, or 0.0002 in.) or less in diameter. The relatively low momentum of the light particles minimizes segregation due to disturbances in the gas flow on entering the nozzle. An interesting case would occur if only very heavy particles were present in the gas stream. Any sampling rate could be used, but the concentration of the particles would be calculated by using the sample volume based on an isokinetic-sampling rate. In actual practice isokinetic-sampling procedures should be followed because the true particle-size distribution of the effluent is not usually known.

Location of the Sample Station

The selection of the sampling-station location has already been discussed. In addition to the points noted a further factor may be added: location of the sample station at a point where gas-flow conditions are uniform allows a representative sample of particulate matter to be withdrawn from a single point in the stack cross section. If this is not possible and it is suspected that segregation of particles is occurring due to gravity effects, varying velocity, or other causes, multiple-point sampling is necessary.

Sample Volume

An accurate determination must be made of the volume of the gas sample in order that concentrations of the components can be calculated. To determine the magnitude of the sample volume that should be withdrawn, several factors must be considered. Chief among these are (a) the sampling rate required for efficient operation of the collection devices, (b) the length of sampling time required to obtain an average representative sample of the emissions, (c) the expected concentration of the specific constituent in the gas stream, (d) the required isokinetic-sampling rate, and (e) the minimum size of particulate-matter sample required for procedures such as particle-size determinations.

Probes and Nozzles

The term "sampling probe" refers to the tube inserted into the gas stream; the term "nozzle" refers to the tube inlet opening for the stack gases.

For most tests one-piece probes and nozzles made of borosilicate glass tubing are used because they afford advantages in size selection, transparency, ease of cleaning, and resistance to corrosion. Temperatures above 1200°F require that quartz composition glass or stainless steel be used. The probe is formed into a smooth short-radius 90° bend, with the nozzle facing upstream into the gases. The straight inlet portion of the probe may be short, except in cases where sampling is done at the top of the stack, when crosswinds may require that the probe extend well down into the stack. Metal nozzles with sharp-edged openings to minimize turbulence are available commercially. Glass probes slightly fire polished on the outer nozzle edge have been found to be satisfactory.

The inside diameter of the nozzle is selected to provide isokinetic-sampling conditions at a volumetric sampling rate necessary for efficient operation of the collection equipment. The methods for holding the probe in the stack (through test holes or at the stack outlets), as well as the rigging of test equipment at the sampling station with the proper relationship of various units to each other, are limited only by the mechanical ingenuity of the source testing engineer.

Sample-Nozzle Selection

By using the actual stack velocity obtained from the Pitot traverse the inside diameter of the sampling nozzle needed for isokinetic sampling is determined from the following equation:

$$d = \left[\left(\frac{R_m}{0.33} \right) \left(\frac{T_s}{T_m} \right) \left(\frac{1}{u_{sp}} \right) \left(\frac{P_m}{P_s} \right) \left(1.02 - \frac{100}{\text{W.V.}} \right) \right]^{\frac{1}{2}}, \quad (10\text{–}12)$$

where d = inside diameter of probe inches,

R_m = flow rate of gases through the flow-measuring device for iso-kinetic-sampling conditions in cubic feet per minute,

T_s = stack-gas temperature in degrees Rankine,

T_m = temperature of the gases passing through the flow-measuring device in degrees Rankine,

u_{sp} = stack gas velocity at the sampling point in feet per second,

P_m = absolute pressure of gas at the meter in inches of mercury,

P_s = absolute pressure of stack gas in inches of mercury,

W.V. = estimated water-vapor content of stack gas in volume per-cent (assuming meter gases contain 2 percent water vapor).

If inside probe diameters are measured in millimeters instead of inches, a value of 0.000507 is used as the constant in (10–12) instead of 0.33.

The known or estimated values of temperature, gas velocity, and water-vapor content are substituted into (10–12), which is then solved for the value of d (assuming for the moment that P_s/P_m equals unity) that will give the desired meter sampling rate R_m. In most cases probe diam-eters fall into the range of 5 to 20 mm ($\frac{3}{16}$ to $\frac{3}{4}$ in.) for meter sampling rates in the region of 0.5 to 1 cfm. The size of the probe increases as the stack-gas velocity decreases. It is usually best for probes to be at least 6-mm ($\frac{1}{4}$ in.) or larger inside diameter to allow for large particles such as fly ash that may be present and also to minimize the relative turbulence due to the leading edge of the sample probe.

After the proper sampling-probe diameter d has been selected, meter sampling rates for isokinetic sampling are calculated for various values of meter pressure P_m that may be expected during the test. This is easily accomplished by multiplying the value for R_m already obtained for zero meter vacuum ($P_s/P_m = 1$) by P_s/P_m ratios for meter vacuums of from 1 to 8 in. Hg. A short table of meter sample rates versus meter vacuum readings is prepared. This will aid adjustment to proper isokinetic-sampling rate during sampling.

Collection or Separation Device

Following the probe is the collection equipment, which may take a variety of forms, depending on the nature of the emissions and the type of data desired. Where wetting of particulate matter is not objectionable a very efficient method makes use of 500-ml size Greenburg-Smith Pyrex-glass impingers, also termed dust-concentration samplers. These are manufactured by the Corning Glass Company as items 6800 or 6820. They are most efficient at sampling rates of about 1 cfm. Three impingers are usually used in series, the first two containing 100 ml each of distilled water. The third impinger is dry and fitted with a thermometer attached to the inside stem to measure the temperature of the outlet gases.

In each impinger the gases are drawn through a small orifice and impinge on a glass plate. The orifice and plate are both below the surface of the water. The high velocity of impingement separates particulate matter from the gas; the particulate matter is retained in the water. In addition, the gases are cooled, and excess moisture is condensed in the impingers. For this purpose the impingers are usually immersed in an ice-water bath. The third impinger further cools the gases and removes water droplets that may have been carried over from the first two impingers. This condensation and cooling also protects the gas meter. Under these conditions the outlet gases are considered to be saturated with water vapor at the temperature indicated by the thermometer inside the third impinger.

Rubber tubing is used for connecting impingers to each other and for subsequent connections to the meter and source of vacuum. Ground-glass connections and adapters may be necessary from the probe to the first impinger, depending on the sample-gas temperature at that point. For sampling of hot stack gases several feet of glass tubing will often serve to cool the sampled gases sufficiently to allow use of a rubber connection to the first impinger. Flexible connections to the impingers are usually desirable because of space or rigging limitations at the sample station.

Where dry filtration is required single-thickness paper extraction thimbles have been found to be efficient dry collectors of particulate matter. One commonly used is manufactured by the Whatman Company, size 43 by 123 mm. The thimble may be mounted inside a glass holder, as shown in Figure 10–7. Metal holders for these paper thimbles may be purchased or fabricated. The gases passing through the paper thimble should be above the dew point, free of moisture droplets, and should

Figure 10–7 Whatman paper thimble and glass holder.

not exceed 250 to 300°F. In a few cases (dust collection from dry gases at close to ambient temperatures) the Whatman thimble may be used alone, but it is usually connected in series with wet impingers. The thimble may precede the impingers (for collection of sulfuric-acid aerosol or for finely divided metal-oxide fumes that are not easily wetted) but is usually placed after the impingers in order to collect any portion of the particulate matter that may have passed through the impingers. Alundum thimbles, size 34 by 100 mm, also may be used for collecting particulates. These thimbles are available in various porosities. Weighings may be made in glass or thin-walled aluminum holders.

Metering Devices

In the great majority of cases, a Sprague dry-gas meter (Zephyr model No. 1A) may be used to meter the gases through the collection train. The meter is usually located on the suction side of the vacuum pump. The collection equipment protects the meter against moisture, high temperatures, and corrosive gases.

The metered gas pressure (relative to atmospheric) and meter temperature are read on a mercury manometer and a thermometer connected to suitable fittings on the inlet and outlet lines, respectively. A trap is located on the inlet line to protect against drawing mercury from the manometer or other foreign matter into the bellows. The meter is integrating in type, similar to a household gas meter, with a large dial reading to hundredths of a cubic foot. The meter flow rate (sampling rate) is determined by difference of dial readings during time intervals. For greater convenience in reading of flow rate (such as in regulation of flow rate for isokinetic sampling), rotameters or orifice meters may be placed in series with the dry-gas meter. The Sprague meter is designed for flow rates as high as 2 cfm. Care must be taken not to subject the gas meter to sudden changes of pressure as the bellows may be injured. These meters should be checked for calibration periodically, depending on their use.

Vacuum Pumps

A vacuum pump suitable for field use for gas sampling trains should be capable of drawing at least 2 cfm of air at vacuums of up to 10 in. Hg or more. A Gast pump may be used to supply vacuum for two separate sampling trains. The Gast pump is suited for long sampling periods. The pump should never be operated without an air filter.

Particulate-Matter Collection

The type of sampling train most frequently used for incinerator testing is shown in Figure 10–8. In it the particulate matter is collected by wet impingement followed by filtration. The sampled gas is passed through three series-connected impingers immersed in an ice-water bath. The first two impingers contain distilled water; the third is dry and fitted with a thermometer. From the impingers the gases pass through a tared Whatman paper thimble, dry-gas meter, and the vacuum pump. Rubber tubing with heavy walls is used for all connections, including the inlet from the probe if the gas temperature is not too high.

Preparation of Equipment

Exactly 100 ml of distilled water is added to each of the first two impingers. The volume must be carefully measured, since later calculations of sample volume depend on an accurate figure for condensate from the stack-gas sample. Extra impingers should be used if very large amounts of condensate are expected.

The Whatman thimbles are tared before the test date by the following procedure. A number of thimbles are suitably marked and heated at 105°C for 30 min and then exposed in open racks to balance room humidity conditions for 24 hr. This process is repeated, and then the thimbles are weighed on an analytical balance to the nearest milligram. The thimbles are stored for use in order of weight. For a test as many thimbles are selected as needed; a standard control thimble is left in the laboratory. This group of thimbles should be fairly close in weight to each other.

All of the equipment is assembled at the test station, and connections are now tested for leaks. The pump and gas meter are run briefly to check

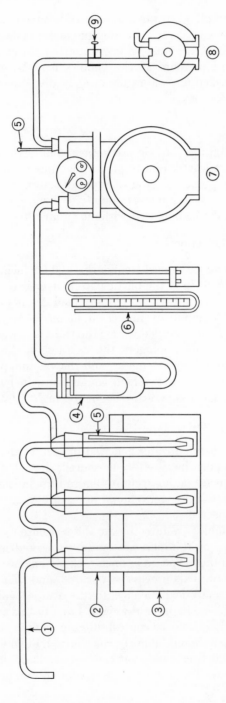

Figure 10–8 Typical incinerator sampling train. 1. Sampling probe. 2. Impinger. 3. Ice—bath container. 4. Dry filter. 5. Thermometer. 6. Mercury manometer. 7. Dry-gas meter. 8. Vacuum pump. 9. Hose clamp to control gas flow rate.

their operation. The sample probes are positioned in the stack, initial gas-meter readings are recorded, and the sampling is started.

Sampling

Data recorded during the test should include the following:

1. Time;
2. Gas-meter reading in cubic feet;
3. Gas-meter vacuum in inches of mercury;
4. Gas-meter temperature in degrees Fahrenheit;
5. Temperature of gases at exit of third impinger in degrees Fahrenheit;
6. Rotameter, or orifice meter flow rate in cubic feet per minute (if used); and
7. Barometric pressure in inches of mercury.

These readings should be taken at 5-min intervals. Total test time is usually 1 hr but may be longer or shorter, depending on the operating cycle of the equipment and the size of sample required.

Notes should be made of any irregularity, such as stoppages of the equipment being tested or of the sampling equipment, during the sampling period. It is obvious that times must be synchronized for multistation tests and when times of operating cycles are being noted.

The sampling rate is regulated to the correct value as soon as possible after the start of the test. This is accomplished in one of three ways, as discussed below, depending on the gas-flow-rate conditions.

Steady Gas Flow Rate—Normal Velocity Profile

The chart of meter sampling rates (cfm) versus meter vacuum (inches of mercury) discussed earlier and prepared previous to the test is now used. The vacuum on the sampling train is regulated by a valve or hose clamp at the vacuum pump until the sampling rate and vacuum correspond to a condition on the chart. For any sampling train with a fixed pressure drop there is only one setting of the regulating valve at which the meter cfm and vacuum are correctly set for isokinetic-sampling conditions at the probe. With practice this adjustment of flow should take only a few minutes at most. Once set, the regulating valve need not be changed during the test unless the pressure drop across the train itself changes. This may be caused by partial plugging of the impinger orifices or of the Whatman thimble by particulate matter. In this case the meter vacuum will rise, necessitating the opening of the regulating valve until alignment on the table is again attained—at higher values of meter cfm and vacuum. This will insure maintenance of isokinetic-sampling conditions.

Frequent readjustment of sample flow rate by the above method is very time consuming and can become inaccurate (because of the time lapse) when attempting to sample from sources with irregular gas flow rates. For this reason an orifice-type flow meter (or a rotameter) is recommended for instantaneous reading of sample flow rate. If the orifice is at the outlet of the vacuum pump, a manometer on the pressure tap of the orifice can then be calibrated to read cfm at atmospheric (stack) pressure. The term P_s/P_m in the general sampling-rate equation drops out (being unity), and R_m becomes a function of V and T_s only, since the other terms are essentially constant. Now all that is necessary is to regulate the pump to maintain the correct sample flow rate as read on the manometer associated with the orifice or as read on a calibrated rotameter.

Steady Gas Flow Rate—Irregular or Distorted Velocity Profile

As previously discussed, multipoint sampling (or dust traversing) is recommended when a very irregular or distorted gas-velocity pattern is shown by velocity traverses at the sampling station. Equal areas of the stack cross section are sampled under isokinetic conditions for equal periods of time. One or more convenient probe diameters are selected, and the various sampling rates are calculated for the different local gas velocities at the point of sampling.

The total sampling time is divided equally between the number of areas being sampled, and the probe is moved to sample each area during the test. Establishment of isokinetic-sampling rate for each sample point can be done by the adjustment of meter rate versus vacuum; use of a flow meter greatly facilitates this operation.

Irregular or Changing Gas Flow Rate

Where irregular or changing gas flow rates are encountered a reference point is selected, and a flue-factor value is calculated from velocity traverses. The Pitot tube and thermocouple are fixed in the stack at the reference-point location prior to sampling. Readings of velocity head and temperature are made at frequent equal intervals, depending on the degree of fluctuation during the sampling period. These intervals may range from 1 to 5 min.

It is found convenient to locate the sample probe as close as possible to the reference point without causing disturbance to the gas flow around the probe and Pitot tube. The changing velocity-head and temperature data from the reference point are used to recalculate and readjust the sampling rate to maintain isokinetic-sampling conditions during the test. To reduce calculations to a minimum tables of sampling rates can be prepared for each sample-probe diameter. These tables give the meter sampling rate in cubic feet per minute at zero vacuum for all expected combinations of velocity head and temperature. The tables are calculated from the general equation, with reasonable assumptions for meter temperature and water-vapor content of the stack gases.

The use of a suitable flow meter to read sampling rate directly is mandatory, because the sampling rate must be adjusted to new conditions of stack-gas velocity and temperature as quickly as possible.

Gas Analysis

Integrated samples of gas for Orsat analysis are collected by a liquid-displacement method. A 5-liter bottle filled with saturated sodium sulfate solution, slightly acidified with sulfuric acid and containing a suitable acid-indicator dye (Figure 10–9) is used as a gas-sample collector. As the

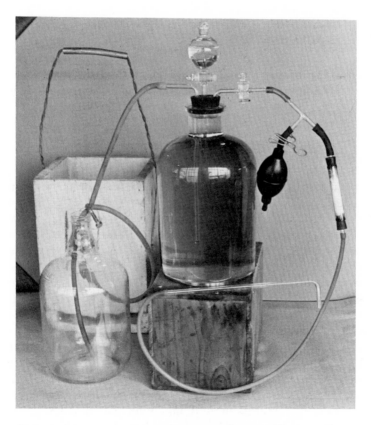

Figure 10–9 Equipment used for collecting integrated gas sample for Orsat analysis.

solution is siphoned out the gas sample is drawn into the bottle. The rate of siphoning is adjusted to give a 3- to 4-liter sample over the test period. Provision is made for flushing the sample line before sampling and for providing slight hydrostatic pressure on the gas sample in the bottle after the test and prior to analysis. This will prevent outside air from being drawn into the bottle.

The pumps are shut off and the trains allowed to come to atmospheric pressure before the vacuum line is disconnected. If the stack is appreciably above or below atmospheric pressure, the inlet lines must be disconnected or clamped before the vacuum line is disconnected to avoid loss of sample.

The final gas-meter reading is recorded. All components of the sampling train that contain particulate matter are suitably stoppered for transfer to the laboratory for processing. Condensate from inlet lines, if any, is allowed to flow into the first impinger.

SAMPLE PROCESSING

The processing of the sample includes two separate operations—one involves determining the quantity of water condensed in the impingers, and the second involves determining the total weight of the particulate matter collected by the sampling train.

Impinger Portion

The total volume of the liquid contained in the impingers is carefully measured. The difference between this volume and the volume of distilled water originally used is recorded as the condensate volume. A more accurate and possibly more convenient procedure is to weigh each impinger before and after the test, making a small correction for the collected particulate matter.

Since the impingers form the first component of the sampling train, they will, in most cases, collect the majority of the particulate matter. To determine the weight of this material the impingers and associated tubing are carefully rinsed with small portions of distilled water, and the liquid and washings are transferred into a beaker or flask. The equipment immediately preceding the impingers, such as the probe and tubing, are also carefully rinsed with distilled water, and the washings are combined with the impinger liquid. Any tarry or organic material in the equipment or tubing should be washed out with minimum amounts of reagent-grade acetone or carbon tetrachloride and added to the aqueous portion. All washing should be done in a counter-current manner (using each portion of water or solvent to successively wash each impinger in a direction opposite to the sample-gas travel) in order to reduce liquid volume and avoid excess usage of organic solvent.

When the gases sampled originally contain no significant amount of the particulate matter that is volatile with steam at 212°F the combined aqueous liquid is transferred to a large beaker, covered with a ribbed watchglass, and evaporated on a hot plate to a volume of about 25 ml. The contents are then quantitatively transferred to a 50- or 100-ml tared beaker, using a rubber policeman to carefully scrub the sides of the large beaker. The small beaker is evaporated just to dryness at 221°F in a constant-temperature electric oven, cooled in a desiccator for 30 min, and weighed on an analytical balance. The difference from the tare weight of the beaker is recorded as the weight of particulate matter collected by the impingers. A blank determination for dissolved solids should be made on each batch of distilled water used and a correction for this blank used for each sample.

When volatile organic particulate matter is collected most if not all of it will be lost if evaporated with water, as described above. In these cases the organic material is removed by extraction with an organic solvent, and the solvent extract is evaporated at room temperature.

To accomplish this the combined aqueous liquid and washings (volume about 350 to 500 ml) are transferred to a separatory funnel and extracted with five 25-ml portions of reagent-grade methyl chloroform or carbon tetrachloride.

The two liquids are allowed to separate as much as possible after each shaking, and care is taken not to include any water in the solvent extract that is drained from the lower portion of the funnel after each extraction. Larger volumes of solvent are used if the aqueous volume is much greater than 500 ml. The aqueous fraction is evaporated at 221°F, and the residue is weighed as already described.

The approximate 125 ml of solvent containing the dissolved organic fraction of the particulate matter is placed in a 250-ml conical flask equipped with a two-holed cork stopper. A short glass outlet tube leads to a vacuum line. An inlet glass tube drawn out to about 1-mm diameter at the tip projects downward to a point just above the surface of the liquid. The vacuum is regulated to draw a jet of air over the surface of the solvent and promote rapid evaporation. The inlet air passes through a large-diameter drying tube filled with a desiccant such as Drierite, followed by a small cotton-wool filter ahead of the solvent flask. The flask should be kept in some type of water bath slightly above room temperature to prevent cooling below room temperature, which would slow the evaporation process. The discharge air from the vacuum pump or aspirator should be hooded to a ventilation system to remove the solvent vapors.

When the solvent has evaporated to 15 ml or less the liquid is transferred to a tared 50-ml beaker, using small amounts of solvent. The beaker is placed under a small bell jar (such as Corning No. 6880) with an arrangement for drawing a stream of dry air over the surface of the liquid at room temperature in the manner described in the preceding paragraph. The evaporation is continued until all of the solvent has evaporated and only an oil or resin remains. The difference from the tare weight represents the weight of solvent-soluble particulate matter collected by the impingers. There should be negligible blank weight from the evaporation of the pure solvent. The weights of the solvent and aqueous residues are added to give the total weight of particulate matter collected by the impingers.

Only relatively high-boiling organic compounds (over 320°F boiling point) will be retained during the evaporation of the chlorinated solvent. The lower boiling organic compounds will not be held and must be determined separately, either by chemical reactions (e.g., aldehydes, ketones, organic acids) or by instrumental methods.

Thimble

The sample thimble or thimbles, together with the control or standard thimble, are heated in an oven at 221°F for 30 min, allowed to stand in an open rack in the balance room for 24 hr, then weighed to the nearest milligram.

The difference in weight of the standard thimble from its original tare weight is used to correct the apparent difference of the sample thimble weight from its tare weight. The net difference is recorded as the weight collected by the sample thimble. The use of a control thimble in this manner serves to correct for differences in moisture content due to atmospheric humidity changes that occur between the time of tare weighing and sample weighing of the paper thimbles.

Gas Sample

Samples of the displacement gases are analyzed by the standard Orsat procedure for carbon dioxide, oxygen, and carbon monoxide, in that order, by absorption in the respective pipets containing potassium hydroxide, sodium pyrogallate, and acid cuprous chloride solutions. Nitrogen is obtained by difference. It is advisable to perform the analyses as soon as possible after sampling. Provision is made for displacement of the gas sample into the Orsat apparatus using the same liquid as for sampling.

The Orsat procedure gives volumetric concentration (to the nearest 0.2 percent by volume) of each component in the stack gas on the dry basis. Calculation to the stack basis may be made by using the stack-gas moisture content.

CALCULATION OF RESULTS

The sequence of calculations from data obtained during sampling and processing of the collection trains for particulate matter is as follows:

1. Volume of sampled stack gas;
2. Water-vapor content of stack gas;
3. Stack-gas density;
4. Particulate-matter concentration;
5. Combustion contaminants;
6. Mass flow rate;
7. Efficiencies.

Volume of Stack-Gas Sample

The total volume of stack-gas sample is equal to the volume of the gas passing through the gas meter in the collection train plus the volume of water vapor condensed from the stack gases in the impinger train. The latter is calculated by the relation

$$V_V = 0.00267 V_L \frac{T_M}{P_M}, \tag{10-13}$$

where V_V = volume of water vapor condensed, in cubic feet at meter temperature and pressure;

V_L = volume of condensate measured in the impinger train in milliliters;

T_M = average gas temperature at meter during sampling in degrees Rankine;

P_M = average gas pressure at meter during sampling, in inches of mercury absolute.

The total volume of the stack-gas sample is calculated at meter conditions as

$$V_{TM} = V_M + V_V, \tag{10-14}$$

where V_{TM} is the total sample volume in cubic feet at meter temperature and pressure, and at standard conditions as

$$V_T = (V_M + V_V) \frac{P_M}{29.9} + \frac{520}{T_M}, \tag{10-15}$$

where V_T is the total sample volume in standard cubic feet at 60°F and 29.9 in. Hg (1 atm) and V_M is the volume of metered gas in cubic feet at meter temperature and pressure.

Examples of these calculations may be found at the end of this chapter.

Water-Vapor Content of Stack Gas

The water-vapor or moisture content of the stack is given by

$$\text{W.V.} = 100 \, \frac{(V_V + V_{VM})}{V_{TM}}, \tag{10–16}$$

where W.V. = water vapor or moisture content in percent by volume;

V_{TM} = total volume sampled, in cubic feet at meter temperature and pressure;

V_{VM} = volume of water vapor in metered gas in cubic feet at meter temperature and pressure;

$$V_{VM} = V_M \frac{(P_{H_2O})}{(P_M)},$$

where P_{H_2O} is the vapor pressure of water in inches of mercury at average impinger exit temperature during sampling.

This relation holds provided that the impinger exit gases are saturated with water vapor and there is no significant change in gas pressure between the impingers and the meter. An example of these calculations may be found at the end of this chapter.

Stack-Gas Density-Correction Factor

When the water-vapor content of the stack gas has been calculated the specific gravity of the stack gas (relative to air) is calculated by summing the individual contributions of the principal components of the stack gas (water, carbon dioxide, carbon monoxide, oxygen, and nitrogen) to obtain an average molecular weight.

The analyses for carbon dioxide, carbon monoxide, oxygen, and nitrogen are converted from volume-percentage dry basis (Orsat) to the stack basis by multiplying the individual percentages by the moisture correction, which is equal to $(1 - \text{W.V.}/100)$. The calculation for each component is then as follows:

$$W_C = (X_C)(\text{M.W.}), \tag{10–17}$$

where W_C = weight of component per mole of stack gas in pounds per pound-made,

X_C = mole fraction of component (equal to volume percent, stack basis $\div$ 100),

M.W. = molecular weight of component in pounds per pound-mole.

The individual W_C values for all components are summed to give $(\text{M.W.})_{av}$, the average molecular weight of the stack gas. The stack-gas specific gravity, D, relative to air is then given by

$$D = \frac{(\text{M.W.})_{av}}{28.95}, \qquad (10\text{--}18)$$

where 28.95 is the molecular weight of air.

A gas-density correction factor, $\sqrt{100/K}$, used to correct the gas velocities from the Pitot measurements, is used since the gas velocity is inversely proportional to the square root of the stack-gas density compared to air.

An example of calculations for the stack-gas density-correction factor is shown at the end of this chapter. This calculation can frequently be omitted for stack gases containing relatively low percentages of water vapor or carbon dioxide.

Particulate-Matter Concentration

The concentration of particulate matter in the stack gas is given by

$$C = 15.43 \frac{(W)}{(V_T)}, \qquad (10\text{--}19)$$

where C = concentration of particulate matter in grains per standard cubic foot, at stack conditions;

W = total weight in grams of particulate matter collected by the components of the sampling train;

V_T = sample volume in standard cubic feet.

An example of this calculation is shown at the end of this chapter. Many government agencies limit the concentration of particulate matter that may be discharged from any source.

Combustion Contaminants

In cases of emissions classed as combustion contaminants concentrations are expressed as grains per standard cubic foot, calculated to 12 percent carbon dioxide at standard conditions. This calculation is performed as follows:

$$C_K = \frac{12C}{(\text{CO}_2)_s}, \qquad (10\text{--}20)$$

where C_K = particulate-matter (combustion-contaminant) concentration in grains per standard cubic foot at 12 percent carbon dioxide;

C = particulate-matter concentration in grains per standard cubic foot, at stack conditions;

$(\text{CO}_2)_s$ = carbon dioxide concentration in volume percent, at stack conditions.

The carbon dioxide concentration at stack conditions is obtained by multiplying the Orsat carbon dioxide analysis (volume percent, dry basis) by the factor $(1.0 - \text{W.V.}/100)$, where W.V. is the stack-gas water-vapor content in percent by volume. An example of this calculation is shown at the end of this chapter.

Mass Flow Rate

The mass flow rate, or rate of emission, of particulate matter at the sampling-station location is given by

$$M = \frac{60CG}{7000} = 0.00857CG, \tag{10-21}$$

where M is the mass flow rate of particulate matter in pounds per hour and G is the stack-gas flow rate in standard cubic feet per minute.

An example of this calculation is shown at the end of this chapter. Many government agencies also limit the hourly weight discharge of dusts and fumes to atmosphere to amounts determined by the magnitude of the process.

Efficiencies

When the inlet and outlet gases to a control system are simultaneously sampled the efficiency of the collection system is obtained by

$$E = \frac{100(M_i - M_o)}{M_i}, \tag{10-22}$$

where E = collection efficiency in percent,
   M_i = mass flow rate at inlet in pounds per hour,
   M_o = mass flow rate at outlet in pounds per hour.

Reporting of Incinerator Test Data

The following is an example of tabular format used in reporting the information obtained from a stack test on a 750-lb/hr multiple-chamber incinerator. Use of the tabular forms facilitates the preparation of data, tables, and calculations; it also provides a clear and logical presentation of the test results.

GAS-VELOCITY DATA

Time	Point	Velocity Head (in. H₂O)	Temperature (°F)	Velocity (ft/sec)	Velocity Head (in. H₂O)	Temperature (°F)	Velocity (ft/sec)	Velocity Head (in. H₂O)	Temperature (°F)	Velocity (ft/sec)
12:30	1	0.05	650	21.6	0.05	650	21.6	0.4	650	19.3
	2	0.05	650	21.6	0.05	650	21.6	0.05	650	21.6
	3	0.5	650	21.6	0.04	650	19.3	0.045	650	20.5
	R	0.035	650	18.1	0.04	650	19.3	0.035	650	18.1
	4	0.025	650	15.3	0.03	650	16.7	0.025	650	15.3
	5	0.02	650	13.7	0.02	650	13.7	0.02	650	13.7
	6	0.01	650	9.7	0.01	650	9.7	0.01	650	9.7
	R	0.04	650	19.3	0.03	650	16.7	0.03	650	16.7
	7	0.01	650	9.7	0.01	650	9.7	0.01	650	9.7
	8	0.02	650	13.7	0.02	650	13.7	0.02	650	13.7
	9	0.025	650	15.3	0.03	650	16.7	0.025	650	15.3
	R	0.035	650	18.1	0.04	650	19.3	0.035	650	18.1
	10	0.05	650	21.6	0.04	650	19.3	0.045	650	20.5
	11	0.05	650	21.6	0.05	650	21.6	0.05	650	21.6
	12	0.05	650	21.6	0.05	650	21.6	0.04	650	19.3
	R	0.04	650	19.3	0.03	650	16.7	0.03	650	16.7
			650	17.2		650	17.1		650	16.7
			Ave.	Average		Ave.	Average		Ave.	Average
	R	Average	650	18.7		650	18.0		650	17.4

A. Indicated velocity (traverse) (ft/sec) __17.0__

B. Indicated velocity (reference point) (ft/sec) __18.0__

C. Flue factor, A/B __0.95__

D. Pitot correction factor __1.0__

E. Gas-density correction factor __1.01__

F. Gas-pressure in stack (in. Hg abs.) __29.9__

G. Gas-pressure correction factor, $\sqrt{29.9/F}$ __1.0__

H. Corrected velocity, $A \times D \times E \times G$ (ft/sec) __17.0__

or $B \times C \times D \times E \times G$ (ft/sec) _____

J. Area of flue (ft²) __4.9__

K. Average flue temperature (°F) __650__

L. Flow rate, $H \times J \times 60$ (cfm) __5000__

M. Flow rate, $(F/29.9) \times 520 \times L/(H + 460)$ (scfm) __2340__

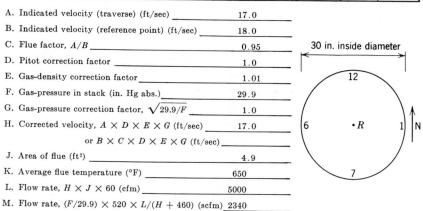

30 in. inside diameter

Table 10–3 Example of Test on Incinerator.

SAMPLING STATION   Incinerator stack outlet   DATE   1–15–68

GAS-VELOCITY DATA

Barometric pressure, in. Hg

Time	Point	Velocity Head (in. H₂O)	Temperature (°F)	Velocity (ft/sec)	Velocity Head (in. H₂O)	Temperature (°F)	Velocity (ft/sec)	Velocity Head (in. H₂O)	Temperature (°F)	Velocity (ft/sec)
1:15	Reference									
+5	"	0.02	750	14.3						
+10	"	0.02	800	14.6						
+15	"	0.02	900	15.2						
+20	"	0.025	1000	17.5						
+25	"	0.02	950	15.5						
+30	"	0.04	1000	22.2						
+35	"	0.035	940	20.3						
+40	"	0.04	980	22.1						
+45	"	0.04	910	21.5						
+50	"	0.03	850	18.2						
+55	"	0.035	970	20.5						
+60	"	0.03	840	18.2						
			910	18.4						
			Ave.	Average						

A. Indicated velocity (traverse) (ft/sec) _____

B. Indicated velocity (reference point) (ft/sec) _____ 18.4 _____

C. Flue factor, A/B _____ 0.95

D. Pitot correction factor _____ 1.0

E. Gas-density correction factor _____ 1.01

F. Gas pressure in stack (in. Hg abs.) _____ 29.9

G. Gas-pressure correction factor, $\sqrt{29.9/F}$ ____ 1.0

H. Corrected velocity, $A \times D \times E \times G$ (ft/sec) _____

       or $B \times C \times D \times E \times G$ (ft/sec) __ 17.7 __

J. Area of flue (ft²) _____ 4.9

K. Average flue temperature, (°F) _____ 910

L. Flow rate, $H \times J \times 60$ (cfm) _____ 5200

M. Flow rate, $(F/29.9) \times 520 \times L/(H + 460)$ (scfm) 1970 _____

30 in. inside diameter

Table 10–3 Continued.

286

SAMPLING-TRAIN DATA AND CALCULATIONS

Time	Gas Meter			Impinger Temperature (°F) T_I
	Reading (ft³) V_M	Vacuum (in. Hg) P_M	Temperature (°F) T_M	
1:15	97.98			
+5	102.5	7.3	100	64
+10	107.0	7.0	101	62
+15	111.8	7.4	102	60
+20	116.2	7.4	106	61
+25	120.0	7.6	108	64
+30	125.4	8.0	112	66
+35	130.3	8.0	115	66
+40	135.0	8.1	119	66
+45	140.0	8.0	120	65
+50	145.0	8.0	120	64
+55	149.7	8.0	118	63
+60	154.71	8.0	117	63
	56.73	7.7 Ave.	111 Ave.	64 Ave.

Material collected   Combustion contaminants

Weight collected (grams):

  Impinger portion 0.145   Sampling point   #3

  Thimble portion 0.059

                                     Sampling-nozzle inside diameter (mm)

                                       Atmospheric pressure

                                                 P_A (in. Hg)   29.9

A. Total weight (grams)   0.204

B. Stack-gas flow rate (scfm)   1970

C. Water vapor condensed, liquid volume (ml)   75

D. Water vapor condensed, vapor volume, $0.0464 \times C$, (scf)   3.48

E. Gas volume metered, $17.4 \times V_M \times (P_A - P_M)/(460 + T_M)$ (scf)   38.4

F. Total gas volume sampled, $D + E$ (scf)   41.9

G. Material concentration, $15.43 \times A/F$ (gr/scf)   0.075

H. Material flow rate, $0.00857 \times B \times G$ (lb/hr)   1.3

I. Material   concentration, $(0.075 \times 12)/5.4$ (gr/scf)
                                  at 12 percent carbon dioxide   0.18

Table 10–3 Continued.

287

TEST NO.   100   PAGE   47

SAMPLING STATION   Incinerator stack outlet   DATE   1–15–68

WATER-VAPOR AND GAS-DENSITY CALCULATIONS

PERCENT WATER VAPOR IN GASES

A. Gas pressure at meter (in. Hg abs.)   22.2

B. Vapor pressure of water at impinger temperature (in. Hg)   0.601

C. Gas volume metered (scf)   38.4

D. Water vapor metered, $C \times B/A$ (scf)   1.04

E. Water vapor condensed, vapor volume, (scf)   3.48

F. Total water vapor in gas sample, $D + E$ (scf)   4.52

G. Total gas volume sampled, $C + E$ (scf)   41.88

H. Percent water vapor in gas sample, $100 \times F/G$ 10.8

GAS-DENSITY CORRECTION FACTOR

Component	Volume Percent/100	$\times$ Moisture Correction $1 - H/100$	$\times$ Molecular Weight	= Weight per Mole of Stack Gas
Water	0.108	1.0	18.0	1.94
Carbon dioxide	0.060 dry basis	0.892	44.0	2.36
Carbon monoxide	0.000 dry basis	0.892	28.0	0.00
Oxygen	0.126 dry basis	0.892	32.0	3.6
Nitrogen and inert gases	0.814 dry basis	0.892	28.2	20.4

J. Molecular weight of stack gas   28.30

K. Density of gas referred to air $= J/28.95 =$ 0.98

L. Gas-density correction factor $= \sqrt{1.00/K} =$ 1.01

Table 10–3   Continued.

TEST NO.   100

SUMMARY OF DATA AND RESULTS

TEST SITE, EQUIPMENT, AND TEST CONDITIONS:
  1. Name of firm   A.B.C. Company
  2. Basic equipment   Multiple-chamber incinerator
  3. Equipment tested   Incinerator
  4. Testing condition and location   Incinerator stack outlet

GAS FLOW AND ANALYSIS:
  5. Gas temperature (°F)   910
  6. Gas velocity (fps)   17.7
  7. Gas flow rate (scfm)   1970
  8. Gas analysis, dry basis (volume percent):

Carbon dioxide	6.0
Oxygen	12.6
Carbon monoxide	0.0
Nitrogen	81.4

  9. Gas analysis, stack basis (volume percent):

Water vapor	10.8
Carbon dioxide	5.4
Oxygen	11.2
Carbon monoxide	0.0
Nitrogen	72.6

AIR CONTAMINANTS MEASURED:
  10. Material collected   Combustion contaminants
  11. Total gas volume sampled (scf)   41.9
  12. Weight collected (grams)   0.204
  13. Material concentration:

Grains per scf	0.075
Grains per scf at 12% CO_2	0.18

  14. Material flow rate (lb/hr)   1.3

Table 10–3 Continued.

INDEX